An Image of God

An Image of God

The Catholic Struggle with Eugenics

SHARON M. LEON

The University of Chicago Press
Chicago and London

Sharon M. Leon is director of public projects at the Roy Rosenzweig Center for History and New Media and research associate professor of history at George Mason University.

The University of Chicago Press, Chicago 60637
The University of Chicago Press, Ltd., London

Printed in the United States of America

22 21 20 19 18 17 16 15 14 13   1 2 3 4 5

ISBN-13: 978-0-226-03898-8 (cloth)
ISBN-13: 978-0-226-03903-9 (e-book)

Library of Congress Cataloging-in-Publication Data

Leon, Sharon Mara.
An image of God : the Catholic struggle with eugenics / Sharon M. Leon.
pages ; cm
Includes bibliographical references and index.
ISBN 978-0-226-03898-8 (cloth : alkaline paper) — ISBN 978-0-226-03903-9 (e-book) 1. Sterilization (Birth control)—Religious aspects—Catholic Church. 2. Catholic Church—United States—History—20th century. 3. Eugenics—United States—History—20th century. 4. Eugenics—Moral and ethical aspects. 5. Religion and science—United States—History—20th century. I. Title.
BX1407.E85L46 2013
231.7′6520973—dc23

2012037997

♾ This paper meets the requirements of ANSI/NISO Z39.48-1992 (Permanence of Paper).

CONTENTS

A NOTE ON SOURCES

In doing the research for this project, I examined a wide variety of print sources. After thoroughly reading the existing literature on the eugenics movement in the United States and the US Catholic history of the period, I embarked on an effort to examine the archival sources that were available both for US Catholic thinkers and for eugenics advocates. These intensive examinations of both Catholic and eugenics sources pointed to the other statements, publications, controversies, and individuals that make appearances in this narrative. I also draw on the official pronouncements of two key entities: the US legislative and judicial system, and the Vatican. On the one hand, both the US Congress and the legislative bodies of the individual states acted on the complex of eugenics policies that worked to define and circumscribe the content and character of the population, assessing the balance of individual rights and the prerogatives of the state, pronouncing who was fit for citizenship, and forming the racial boundaries of the nation. At times, the US judicial system offered confirmation or condemnation of these legislative judgments. On the other hand, the Vatican produced a body of social and moral teaching based on the traditional teachings of the Roman Catholic Church on individual rights, fitness, and race that formed the basis for Catholic responses to the eugenics movement. In some sense, then, these two entities provided the flexible but implicit boundaries for the conversations that this book analyzes, suggesting limits to the eugenics agenda and to the range of Catholic responses.

To uncover the Catholic perspective, I began with an examination of the institutional records of the National Catholic Welfare Conference (NCWC), which are housed at the Catholic University of America's American Catholic History Research Center and University Archives. Though known by a number of names in the twentieth century, the NCWC served as the central organization for the American Catholic hierarchy after World War I. If eugenics was something about which the church as a whole needed to be concerned, some conversation and action would have taken place in those offices. Within the NCWC papers, materials from the Office of the General Secretary, the Social Action Department, the Lay Organization Department, the Legal Department, the National Council of Catholic Men, the National Council of Catholic Women, and the Family Life Bureau were essential to reconstructing the Catholic response to eugenics initiatives. The NCWC papers then led me to explore the personal papers of both the Reverend John A. Ryan and the Reverend John Montgomery Cooper, prominent priests and members of the faculty at the Catholic University of America who were interested in and involved with the eugenics movement. To follow up on this archival work, I did a survey of the available Catholic periodical literature that turned up a good deal of material from both ecclesiastical journals and more popular Catholic reviews. This research was made a great deal easier by the existence of the *Catholic Periodical Index*, which provides a comprehensive keyword index of the contents of major and minor Catholic magazines beginning in 1930.[1] I hand-surveyed material prior to that date using the indexes of individual publications such as *Commonweal, America, Catholic World, The Catholic Mind*, the *American Ecclesiastical Review*, and the *Homiletic and Pastoral Review*.

From this point, I turned my focus to the local level. I queried diocesan archivists in areas where there had been eugenics activity. For the most part, unfortunately, those archives yielded little. The most promising materials resided at the Cleveland Diocese, due to Bishop Joseph Schremb's oversight of lay organizations and his heightened awareness of eugenics policies. Though the diocesan archival material was slim, diocesan newspapers yielded more material. Key sites included Alabama and California. Again, I based my survey on locations with strong eugenics organizations and/or legislative action.

In addition to pursuing the Catholic voices in this conversation, I investigated the range of mainline eugenics materials for points of contact, conflict, and consensus with American Catholics. First, I reviewed the full range of books dedicated to eugenics that were published between 1900 and 1950 in the United States. From this base, I moved to the papers of the American Eugenics Society and the papers of major figures in the movement, including Charles Davenport and Fredrick Osborn, which are housed at the American Philosophical Society. Second, I read the full run of both of the AES publications, *Eugenics* (1928–31) and *Eugenical News*

(1916–53). These two periodicals consistently included articles that exposed the conflict between eugenics ideas and Catholic teaching and actions. Additionally, *Eugenics* hosted a number of roundtable debates that featured prominent Catholic thinkers, allowing for direct comparison of viewpoints on such key issues as immigration restriction, the value of charity, racial discrimination, and birth control.

AN IMAGE OF GOD, NOT A MERE ANIMAL: AN INTRODUCTION

On May 2, 1927, the Supreme Court issued its decision in *Buck v. Bell*, which has become the most famous decision in the history of the American eugenics movement. Judged "feeble-minded" by a commission of experts and the superintendent of the State Colony for Epileptics and Feeble-Minded in Lynchburg, Virginia, Carrie Buck—like thousands of institutionalized persons before and after her—had been in 1924 ordered to be sterilized under the authority of the state. The Supreme Court, in an 8–1 vote, affirmed the state's actions, thus demonstrating the power of eugenic thinking in the United States. In October 1927, Carrie went under the knife, and she was released in November.

For most of us, eugenics is now a distant memory. The concept—that science can improve a population through selective breeding, and that controlling heredity is the key to reforming society—seems almost absurd. Thanks to the horrors of Nazi atrocities during World War II, clearly influenced by eugenic ideology, the organized eugenics movement has been thoroughly discredited. In contemporary America, public policy and popular culture alike seem (at least most of the time) to be built on far more liberal, humane ideals. Most of us, it seems, believe that we, as individuals, have the right to live our own lives as we see fit, and that the state does not have the right to decide whether, or how, we reproduce. And yet the vast change in consciousness over the last century can obscure the ongoing relevance of eugenic thinking. The movement is far from a relic. Instead, as we shall see, the eugenics movement, and those who opposed it, remains essential for our understanding of America, past and present.

In the half-century before World War II, the idea of eugenics captivated a broad swath of the American public. The movement drew momentum from disparate sources: from the progressive impulse to systematize and control reform; from popular alarm about an influx of working-class immigrant newcomers; and from concern about the upheavals created by urbanization and industrialization. Keywords such as "imbecile," "defective," and "unfit" were code for individuals who, in varying degrees, failed to fulfill society's desired notions of middle-class productivity and respectability. Eugenicists held that bad heredity was to blame for such failures; society's blights, according to eugenics logic, could be halted by discouraging reproduction in those individuals. Conversely, the "fit" and "superior" among us could and should be encouraged to multiply. Eugenic ideology was transformed easily into action: proponents wrote reams of articles and advice manuals advocating "proper" lifestyle choices; created programs to urge the appropriate citizens to reproduce; and fought for public policies promoting everything from immigration restriction to involuntary sterilization, from health requirements for marriage certification to stricter antimiscegenation standards.

Affirming the Commonwealth of Virginia's involuntary sterilization law, which provided a way for superintendents at state hospitals and institutions to have inmates sterilized against their will, Justice Oliver Wendell Holmes Jr., in the majority opinion for the *Buck* case, penned the oft-quoted sentence: "Three generations of imbeciles are enough." Holmes's reasoning was viciously clear: "We have seen more than once that the public welfare may call upon the best citizens for their lives. It would be strange if it could not call upon those who already sap the strength of the State for these lesser sacrifices, often not felt to be such by those concerned, in order to prevent our being swamped with incompetence. It is better for all the world, if instead of waiting to execute degenerate offspring for crime, or to let them starve for their imbecility, society can prevent those who are manifestly unfit from continuing their kind."[1] Though surprising in our own, more diverse public culture, the honesty of his vitriol surprised few in 1927. Such was the influence of the eugenics movement.

But a lack of surprise does not imply total agreement. The *Buck* decision, like every other victory for supporters of eugenics, prompted negative responses from various segments of American society. No denunciation was more forceful than the editorial that ran in the Jesuit journal *America* on May 14, 1927. The editors at *America* did not deny that the state had the right to sterilize the "mentally defective" to protect the common good under certain circumstances (we shall see that there were some troubling similarities of thought among opponents and proponents of eugenics).

But on the whole the editors maintained that the *Buck* case did not necessitate sterilization, and that there were a number of less-invasive alternatives available to the authorities in Virginia. They argued that sterilization was a shortcut to social health, a shortcut that would lead to a further dissolution of American society. Calling for a focus on individual integrity, they explained: "Fundamentally, our objection is based on the fact that every man, even a lunatic, is an image of God, not a mere animal, that he is a human being, and not a mere social factor. To care for the dependent with sacrifice, foresight and charity, is a work which ennobles the individual, and is a source of vigor for the State. To care for them with a surgeon's knife and nothing else and then to stamp this method as 'enlightened' shows how far we have wandered from the concepts of humanity and of Christian civilization."[2]

The potency of this argument is revealing. As we shall see, Americans from across the religious and political spectrum disagreed with eugenics. Not surprisingly, academics were in the forefront of disagreement. The most publicly vocal of these included the anthropologist Franz Boas and his students, who embraced "culture" as a better explanation for difference than "race."[3] Similarly, individual sociologists stepped forward to reject the basic premises of eugenic thinking—including Lester Frank Ward and Miami University's Warren Thompson, who claimed that standards regarding intelligence reflected the prejudices of elites.[4] Outside of the academic community, conservative evangelical Christians such as Robert Speer and Walter Maier expressed some opposition to hard-line eugenics, but their voices were few and far between among Protestant clergy.[5] These individuals, and many others, offered critiques of the movement, some of which were very influential.

But across the nearly half-century of eugenics successes, no group was as successful in their opposition as American Catholics. Academics and other individuals could not claim to speak for large constituencies; the Catholic hierarchy, on the hand, did just that, opposing eugenic thinking and policies in the name of the church and its millions of parishioners. Time and again, as we shall see, members of the Catholic hierarchy, as well as priests, physicians, writers, and organized lay men and women, spoke out in the press, within their own organizations, and in the legislative arena against the invasive and restrictive measures inspired by eugenics. Thus, Catholics comprised the most significant and organized opposition to eugenics policies. This is the story of the clash between Catholics and eugenicists across four decades and its numerous ramifications for how we understand American society, the relationship between church and state, individual rights, and politics.

Placed side by side, Holmes's opinion in the *Buck* case and the *America* article in response reveal a great deal about the dynamics of Catholic encounters with the eugenics movement in the United States during the first half of the twentieth century. In both perspectives, there is concern for the common good, for the obligations of the community, and for the rights of the individual. Additionally, in both cases there is an agreement that the "mentally defective" should not have children. But it is unclear whether the *America* editors believed that such disability rendered an individual incapable of consensual and responsible parenthood, or whether, as in the opinion of Justice Holmes, they believed that mental disability is hereditary. It is quickly clear that these debates bleed into fundamental issues of human life: the obligations of the state; the place of heredity in determining behavior; and the rights of the individual to bodily integrity and to reproduction.

Despite their commonalities, the conclusions that the authors reach are dramatically different. Holmes suggests that "defective" individuals should be willing to sacrifice for the common good, whereas the editors at *America* argue that the community has a duty to sacrifice, to care for the less able, and in fact that such sacrifice contributes to the overall health of the community. The editors of *America* and other Catholic opponents of eugenics repeatedly called for the protection of individual rights within the context of a strong community—in this case, the state—that recognizes the potential for every member to contribute to the common good, regardless of material limitations, and that recognizes an obligation to charity. This perspective is in contrast to the eugenic view that a person's biology constitutes an important factor, and often the primary factor, in determining his or her value to the community.

This fundamental dispute, over whether the state has a right to sacrifice an individual for the common good or whether the state has the obligation to care for all individuals as an integral part of the common good, provided the basis of the conflict between proponents of eugenically inspired policies and Catholic opponents of those policies. While supporters of eugenic policies saw hereditary science as the key to transforming American society—because it provided a guide for preventing both the physically and mentally "unfit" from reproducing—their Catholic opponents were much more focused on environmental and religious sources of social improvement, and disdained proposals that called for the state to circumscribe an individual's natural right to marry and have children. They believed that individuals could not be defined solely by their biology.

I argue that Catholic clergy, writers, and activists urged their co-religionists to object to eugenics for a number of key reasons. First, they believed that the science that eugenicists used was not rigorous enough to justify their policies. Rather, those scientific claims were simply a guise for more deeply held ethnic, racial, and class prejudices. Second, they rejected the notion that the state and the community that it represents can ever justly assume the power to violate the bodily integrity of an innocent individual, regardless of the supposed biological improvements that will result. Catholics did not disagree with the notion that humans should work to improve the health and welfare of the population, but they did disagree that the goal of biological improvement superseded all other rights, values, and goals. For nearly forty years, Catholic thinkers and eugenics supporters argued about these ideas.

In a book about how and why American Catholics opposed some of the ideas and many of the initiatives of the eugenics movement, one might assume a narrative that rehearses the tired story of the "war" between science and religion. For the most part, however, that age-old trope bears little resemblance to the reality of this interaction. Rather, Catholics addressed eugenics on the grounds that the science was inadequate or poorly done and that the measures that the movement called for were not the best methods to safeguard the rights of individuals while ensuring the common good. More often than not, Catholic thinkers objected to the social application of scientific findings and principles rather than to the science itself. In fact, Catholics made their arguments within the grounds of biology, psychology, sociology, and law, rather than on the basis of religious principles. America's religious pluralism, they recognized, necessitated convincing those who did not share their religious convictions that eugenic ideas and initiatives were not in the best interests of American society at large.

Reforming and improving that society was a pulsing current in America's public thought and policy when eugenics achieved its initial burst of popularity in the 1910s. The Progressives, who were prominent in local and state politics in the early years of the twentieth century, championed measures to clean up and modernize the nation's cities, local governments, and public institutions. Many of these reforms focused on environmental conditions—everything from improving the sanitation in urban slums, to regulating food and drug production, to limiting child labor. Improving the world of the poor—their living conditions, their opportunities, their health, so the logic went—would slowly improve America as a whole.

Supporters of eugenics, however, diverged from the majority of Progressive reformers, believing that environmental reforms could not improve the lot of the inferior. While eugenics sympathizers were not all strict hereditarians, the heart of the movement was convinced that biology trumped environmental effects. Only by controlling the basic germ plasma—the genetic makeup—of the population could they transform America. Eugenic reformers believed they could resolve a host of social problems that Progressive efforts to alter the environment had failed to eliminate. Rather than advocating minimum-wage laws, better urban housing, food processing regulations, and education, eugenic reformers were convinced that by controlling reproduction they could eradicate social problems at their root. They could halt social ills before they happened. At the heart of this national re-imagining lay a dual task: halting the reproduction of the unfit and increasing the reproduction of the fit.[6]

The daring aims of the eugenics movement attracted a wide range of followers. As we shall see, there was never one monolithic movement, but rather a spectrum of voices. The "mainline" eugenics movement, as it is often called, was the most well established. It was led by the Eugenics Society of America—later, the American Eugenics Society (AES)—founded in 1921. The typical member of the AES was a middle-aged, upper-middle-class, native-born white man. Despite the typical profile of an AES member, eugenics appealed to a broad range of activists on the left and the right, including a range of feminist activists, who stirred both devotion from their followers and outrage from their detractors. According to the geneticist Raymond Pearl in 1908, eugenics, "to an extraordinary degree," appealed to "radicals and conservatives alike, as something for which the time is quite right."[7] Taken as a whole, the advocates, supporters, and devotees of eugenics came from nearly all walks of white America, manifesting a diversity that makes their ideas that much harder to ignore.

The early movement's base was established by Charles Davenport, known as the father of American eugenics, and Harry Laughlin, his collaborator at the Cold Spring Harbor Station for Experimental Evolution and at the Eugenics Records Office. From those beginnings, the movement gained a popular following in the years before World War I, but it truly achieved its apex of organizational and legislative success during the considerably more cautious and conservative 1920s. The specificity of the institutional eugenics movement's plan for social transformation was dramatically articulated in the "Ultimate Program" of the AES—as proposed in 1923. The proposal's introduction forthrightly stated: "The general aim of the society is to forward the practical application of eugenic principles to the improvement of the American population."[8] In support of that aim, the Society would promote eugenics research, eugenics education, and conservative legislation, including support for mothers' pensions, eugenics education boards, and mental testing regulations as well as immigration restrictions and sterilization laws.[9] The breadth and variety of the measures proposed by the AES covered a whole range of factors touching on all elements of society, promoting the reproduction of superior individuals and discouraging the reproduction of inferior individuals. In varying degrees, this program of social transformation guided the work and activism of eugenics proponents through the 1930s. And though popular enthusiasm for negative eugenic measures waned in light of the Nazi atrocities, the principles of positive eugenics infused postwar pronatalism—the effort to support and encourage large families.

The rationale for their ambitious plan of social control was often quite simple. Prominent eugenics advocates, for example, frequently argued explicitly that their ideas would produce financial gains for the state. If America followed their advice, the state would have huge resources at its disposal—the funds no longer needed to finance charitable programs, prisons, or insane asylums. To Davenport, the task was simple, urgent, and self-evident: "It is a reproach to our intelligence that we as a people, proud in other respects of our control of nature, should have to support about a half a million insane, feeble-minded, epileptic, blind and deaf, 80,000 prisoners and 100,000 paupers at a cost of over 100 million dollars per year."[10] Not only would eugenics reforms free American society of the burden of inferior populations with their concomitant social ills, but citizens would also reap substantial material benefits.

Saving money was merely one of many arguments eugenicists made to enlist the support of politicians, but it was not the eugenicists' main goal. Ultimately, American eugenicists wanted to turn their country into the racially superior nation they knew it had the potential to become. The core supporters of the movement believed that race was a biological reality. The vehemence of this belief is hard to comprehend today, as in all but the most extremist circles the notion of "biological reality" has now been thoroughly discredited by the notion of race as socially constructed. But the fixed reality of race seemed as obvious in the first half of the twentieth century as the color of one's skin. The legislative policies, educational programs, and reams of popular advice created by eugenicists created a "racial project" which was "simultaneously an interpretation, representation, or explanation of racial dynamics, and an effort to reorganize and redistribute resources along particular racial lines."[11] In other words, eugenicists dreamed of—and believed that, with the right support, they could create—an America dominated by people of Northern European heritage and culture, an America that was intelligent, ruddy-cheeked, blond, and hearty.

Deeply entrenched in the early-twentieth-century understanding of race, eugenicists predominantly targeted "white" persons in their policies. While many supporters had lots to say about the degeneracy of African Americans, or the deleterious influence of Asian immigrants, for the most part they felt that there were strong enough social conventions to maintain segregation. Certain historians contend that eugenics was not concerned with race, and instead argue that economics or gender roles were of primary concern. Such arguments, however, project backward late-twentieth-century notions of racial categorization reflected in the five principal divisions of the US census. But the categories themselves—"American Indian or Alaska Native," "Asian," "Black or African American," "Native Hawaiian or Other Pacific Islander," and "White"—create a significantly simplified racial landscape.[12] Eugenicists, and most other early-twentieth-century Americans, thought, wrote, and lobbied for policy measures in a time when, as Matthew Frye Jacobson deftly illustrates, whiteness itself was fractured into a "hierarchy of plural and scientifically determined white races." During this period of increased European immigration, not all persons of European descent saw themselves, or were seen by others, as part of a "unitary" whiteness. The gradations of whiteness—a hierarchy typically inhabited at the top by those of Northern European origin and then followed by the many variations of other immigrants—were exactly what eugenicists wished to control and contain with their restrictive policy initiatives.[13]

As we have already seen, the law was central to eugenics efforts to control the content and character of the US population. The race theorist Ian Haney López describes how legislatures and courts serve "not only to fix the boundaries of race in the forms we recognize today, but also to define the content of racial identities and to specify their relative privilege or disadvantage in US society."[14] Eugenicists championed such legislative and judicial solutions—immigration restriction, compulsory sterilization, and marriage restrictions, to name a few—on the grounds that such measures would ensure racial purity and improvement.

In addition to negative eugenics measures, eugenics proponents promoted "positive" eugenics: efforts to encourage persons exhibiting "superior" hereditary traits to have an adequate number of children. Eugenics proponents usually suggested that each family of superior racial quality needed to have at least four children, both to replace themselves and to increase the population of good stock. "Fitter Families" contests and advice concerning good prenatal care are just two of the methods local eugenics groups used to promote beneficial reproduction.[15] The ideas supporting positive eugenics fit easily into a whole host of reform activities and causes, and were yet another way that eugenics meshed with broader calls for social improvement.

It is important to mention that some aspects of positive eugenics actually appealed to Catholics, because they encouraged married couples and society as a whole to accept and support large families. While Catholics may have rejected the standards by which eugenics supporters decided who they were going to encourage to reproduce, they nonetheless agreed with measures that affirmed family life. In this we see the primary instance where our story is complicated by the similarities between these two groups.

As we shall see, the responses of Catholic thinkers, writers, and activists to eugenics ideas and policies offer a revealing look at both the evolution of Catholicism and of American society at large. These responses were not simply knee-jerk condemnations. Rather, they were far more nuanced, taking into account the traditional teachings of the church, the proper relationship between church and state, the social and material conditions of the Catholic population and the greater American community, the quality of the scientific work forming the basis for eugenics claims, and the underlying assumptions of the activists making those claims. All of these factors resulted in the articulation of a complex body of reasoning and work that reveals Catholics' struggles to participate in modern American social and political life while maintaining their commitment to the church's traditional teachings on race, gender, family, economics, and community. In many cases, more-established, often middle-class, laypersons and clergy aimed to protect their poor and working-class co-religionists, who were more vulnerable to eugenics policy, while at the same time articulating principles and positions that served to protect the poor, the infirm, and the disadvantaged in the population at large.

While providing this bulwark against the invasive policies called for by the eugenics movement, these vocal Catholic activists participated in an even larger effort: to adapt Catholicism to American conditions. The questions that eugenics raised about the social order, race, gender roles, and reproductive restrictions were truly modern ones. Asserting their rights as citizens to participate in the public debate about the importance of environment and heredity, and about the proper role of the state and the responsibility of the individual, Catholics transformed public discourse. These men and women brought to bear the lessons of Catholic moral and social teaching, and their own concrete experience as a diverse community, on the ongoing conversation about the balance between individual rights and the interests of the community. In doing so, they transformed themselves from religious outsiders into an integral and increasingly accepted part of the American community.

The innumerable groups that make up the American community are our greatest strength, and Catholics have been a significant and distinct voice in that community since the nineteenth century. But as a Catholic, I have noticed throughout my life the ambivalent place of my community within American history and consciousness. Writing this book has been in part an effort to understand that ambivalence by probing the way that Catholics acted on their moral teachings in the context of American public life. In telling the story of American Catholic responses to eugenics ideas and policies, this book takes up the call issued by Leslie Woodcock Tentler nearly twenty years ago, when Tentler challenged US historians, arguing that Catholics needed to be "integrated, more fully and intelligently than heretofore, into our reconstruction of the past."[16] The successes and failures of the eugenics movement, as we will see, cannot be fully understood without considering the role of Catholics in publicly debating and often fighting the eugenicists' theories and initiatives. The story of four decades of interaction between Catholics and eugenicists thus illuminates both the history of Catholicism and the history of eugenics in the United States.

In the chapters to come I will discuss how and why Catholics became the major opponents of American eugenics. I will detail the ongoing conflict, from about 1910 to 1950, between the leading figures both from the Roman Catholic community and from the mainline eugenics movement. Our story emerges from the Catholic discourse around the eugenics movement and reveals the thinking of members of the Catholic hierarchy, moral theologians, and prominent members of the clergy and the thin layer of middle-class lay leadership. Hence, there are few voices here of average, working-class, lay Catholics, and few voices of the targets of eugenics policies and practices, though the rights and welfare of these individuals were of major concern to the figures who do find voice in this narrative. As we will discover, the Catholics who do appear in the pages of this narrative did more than engage in intellectual debate with eugenics advocates on how to determine human worth and how to shape the American community. They took action lobbying against eugenics policy initiatives—action that significantly hindered the eugenics movement's legislative agenda. Thus, the book illuminates the ways in which Catholic thinkers and activists, imbued with social and cultural sensibilities shaped by their religious commitment, transformed the American public conversation about individual rights, the role of the state, and the intersection between racial thinking and ideals of family and community.

ONE

For the Welfare of the Race: The Early Clash over Reproduction and Community

In March 1910, Father Stephen M. Donovan wanted to talk about vasectomies. Specifically, he was concerned about the increasing number of vasectomies ordered by an increasing number of states for institutionalized persons. Were these state-ordered deprivations of the ability to have children moral? Writing in the *American Ecclesiastical Review* (*AER*), he launched a discussion that would draw numerous contributions and opinions from Catholic moral theologians in the United States and Europe for the next two years.[1] Rather than addressing lay people, the *AER* provided instruction for the clergy, often advising them on matters of doctrine and morality so that they could more appropriately deal with the challenges of their congregations. A professor at the Franciscan House of Studies affiliated with the Catholic University of America, Donovan initially questioned the morality of the sterilization operation as a practical matter, since a number of states had begun to authorize the practice in order to lessen criminal tendencies and to prevent what they deemed "hereditary degeneration." His inquiry marked the first mention of eugenic sterilization within the pages of an American Catholic publication. The ensuing discussion about the permissibility of the vasectomy operation would result in no definitive pronouncements, but it served as a forum to lay bare some of the conflicts and dilemmas that would recur again and again in the next two decades as US Catholics debated the many facets of sterilization and the larger questions of eugenic reform.

Donovan's interest in the issue coincided with the public recognition of Dr. Harry C. Sharp's use of vasectomy among institutionalized persons in Indiana. The surgeon for the Indiana Reformatory, Sharp had moved to the forefront of those advocating eugenic sterilization with a March 1902 article in which he argued that states should authorize their mental institutions to sterilize inmates. His rationale: the positive and calming effect that he observed from the forty-two inmates on whom he had performed the operation in the previous several years. In undertaking this remedy, Sharp had put into practice a procedure that had only appeared in the medical literature for the first time in 1897. But in the years immediately after publication of his article, others warmed to Sharp's perspective; in 1907 the Indiana legislature affirmed his position by passing the country's first sterilization statute.[2] The law called for the prevention of "procreation of confirmed criminals, idiots, imbeciles, and rapists," and clearly stated the importance of hereditary thinking to this legislative movement—and the budding American eugenics movement—by professing, "Heredity plays a most important part in the transmission of crime, idiocy, and imbecility."[3]

The clear association between heredity and crime can be tied to nineteenth-century criminal anthropology and a host of "family studies" that raised the question of criminal fecundity and the resultant drain on community resources. In 1909, the *Journal of the American Medical Association* published a speech, delivered at the association's annual convention, in which Sharp directly addressed the rapidly rising cost of institutionalization and proposed that the results of his successful sterilization of inmates—now nearly five hundred in all—offered a viable solution. Activist physicians in New York, New Jersey, Virginia, California, Maryland, Illinois, and Texas followed Sharp's lead and called for eugenic sterilization as a solution to the problem of state institutions overflowing with male inmates. By 1913, eleven other states—including Washington, California, Connecticut, Nevada, Iowa, New Jersey, New York, North Dakota, Michigan, Kansas, and Wisconsin—had put eugenic sterilization measures into place. The legislatures of four other states—Pennsylvania, Oregon, Vermont, and Nebraska—passed similar laws, which were then vetoed by their governors.[4]

It was in the context of this growing medical literature and bustling legislative activity that Donovan's piece prompted further inquiries. Physicians, lawyers, and clergy alike wondered about the "attitude which a Catholic is in conscience bound to take should the subject be brought before them in a practical way."[5] In general, the theologians who responded fell into two categories. The first group maintained that vasectomy operations were immoral since the procedure constituted a grave mutilation of the human body and only grave mutilations that were necessary for the conservation of human life could be morally justified. The second group of theologians submitted that, since the patients in question were not only defectives but also criminals, the state had the right to perform the operation for the greater protection of society. Also, since the operation could be reversed upon rehabilitation, the mutilation would not necessarily be permanent.[6] Though the authors each brought particular nuances and concerns to their writing, these two general positions laid out the key issue under consideration in the exchange: under what circumstances, if any, would the state have the right to physically interfere with an individual's capacity to have children.

In exhibiting these contrary positions, the writers in the *AER* began the first in what eventually would be a much larger Catholic exchange about the eugenics movement and the policy initiatives that the movement advocated. While the *AER* contributors limited their reflections to the narrow question of whether the state could justifiably sterilize a criminal inmate, other Catholic thinkers tied eugenics reform theories and goals to a host of other issues, including the appropriate role of science in social policy, the challenges of the birth control movement, and larger questions of economic justice. It is this broad gamut of Catholic thought that shapes the story presented here. For the emergence of the eugenics movement itself was matched, nearly as quickly, with a deep engagement by Catholics with the benefits and pitfalls of that movement. Here, I begin by exploring the variety of ways that Catholic thinkers agreed and disagreed with the claims of the eugenics movement and what these arguments reveal about the proper relationship between the state and the family and the obligations of a society to care for all of its members.

In 1910, however, no one could have anticipated the roiling debates of the next three and a half decades, nor the many corners of ordinary life that would be touched by eugenics. The different perspectives published in the *AER* in response to Donovan's initial query produced no consensus about the morality of state-sanctioned vasectomy for criminals, so in May 1911 the editors turned to P. A. Schmitt, a Jesuit from Louvain University in Innsbruck. Schmitt's response, translated and summarized for *AER*'s American readers, suggested that the morality of vasectomy could only be judged with respect to the intentions of the individual performing the operation. He hoped that this emphasis on context would make the question regarding the double effect of the operation more clear. Schmitt then argued that the serious effect of the operation—withdrawing the powers of reproduction from the patient—yielded an intended good result: preventing hereditary disease, something that "all will agree . . . is desirable in itself." There was a crucial caveat, however. Schmitt held that the good result was not the *immediate* effect of the operation, but only a by-product. In this sense, judging a vasectomy to be morally permissible would be equivalent to arguing that the ends justify the means. In summarizing Schmitt's opinion, the editors reminded their readers, "It is never lawful to do evil in order that good may result therefrom."[7]

In addition to his perspective on ends and means, Schmitt recognized that if the legislation in question were to be logically applied, it would affect more than just mentally defective criminals. It would also apply to those who were "affected by tuberculosis, excessive alcoholism, the sexually morbid in certain stages of disease, and indeed to many other conditions of life in which heredity becomes the immediate source of mental or physical defects cumbersome or dangerous to the commonwealth." Based on this observation, Schmitt made a strong argument that the impulse to sterilize was propelled by a misguided notion that heredity was the primary cause of degeneracy. He strongly disagreed, submitting that "evil has many sources, such as a false method of education, mistaken or insufficient nutrition, especially the use of alcoholic stimulants, social and labor conditions which over tax the capacity of the individual, the strenuosity and nervous anxiety incident to the striving after material success, artificial living and the pursuit of enervating pleasures,—all of which contribute to the growth of insanity or mental and physical degeneracy." Thus, Schmitt surmised that by emphasizing the hereditary causes for degeneracy, reformers absolved themselves from advocating expensive and complex social and environmental reforms that would really help to end degeneracy.[8]

In subsequent months, as letters poured in to the journal, opinions remained evenly split between those who felt that vasectomy was immoral under all circumstances (other than when necessary to preserve the life of the patient) and those who felt that vasectomy served a greater social good which justified the operation. Since the Vatican had offered no explicit teaching on the question, these authors were free to disagree on the morality of state-ordered vasectomies for criminals. No one, however, denied the fact that the legislation as written—giving institutional administrations complete authorization to choose whom to sterilize—would be ripe for abuse. In the end, Stephen Donovan explained:

> While am I still thoroughly convinced of its correctness theoretically, I cannot fail to recognize the wide difference that exists between theory and practice; nor have I at any time during the controversy lost sight of the numerous and great abuses that would follow were the lawfulness of the sterilization of criminals be generally conceded. Practically, therefore, and, as a general rule, the affirmative opinion should not only not be urged in practice, but Catholic surgeons and lawyers as well as others who have an active part in the framing of the laws of our several States should take a firm stand against

> the legalizing of any surgical operation by which criminals are deprived of their procreative faculty.[9]

At its heart, the discussion about whether or not the vasectomy operation violated the Catholic interpretation of moral law laid out three issues at the core of Catholic responses to negative eugenics policy initiatives during subsequent decades. First, all of the theologians addressed the importance of the generative faculty. They agreed that human beings possess a reproductive faculty endowed by the Creator and that individuals are obligated to control and exercise in light of reason. Second, the authors involved in this conversation agreed that the role of the state in controlling and regulating reproduction was a question that theologians and philosophers would have to address before they could come to any consensus on sterilization. The contributors to the *AER* were unable to agree on whether the interests of the community, embodied in the state, superseded the right of the individual to procreate. Finally, the authors suggested that the current sterilization legislation was quite ambiguous about who would be subject to the operation. Thus, there was a possibility for grave abuse and tremendous moral hazard.

As we know, these questions—of reproductive rights, the power of the state, and the rights of the individual—were deeply embedded in the larger eugenics movement, of which the flurry of state sterilization statutes was only a part. Eugenics as a movement grew out of an impulse to discover predictable natural laws that could be applied to improve modern society. The British statistician and traveler Sir Francis Galton (1822–1911) led the way, drawing on contemporary thinking about evolution and the urban-industrial climate to develop a theory about individual greatness and British society in the 1860s. A cousin of Charles Darwin, Galton applied mathematical techniques to the characteristics he observed in the natural world, working with statistical notions of normal distribution illustrated by the bell curve. He began collecting human data and pondering questions of human heredity using mathematical expressions for the relationships between measured traits. Tracing the family history of prominent individuals—namely scions of the British upper class—in "Who's Who" guides such as the *Dictionary of Men of the Time*, Galton claimed to have proven a hereditary cause for "greatness" and success. Thus, he posited that the race could be improved with a program of eugenics, or "good breeding"—purposefully encouraging the "fit" to reproduce and discouraging the "unfit" from reproducing.[10] Galton's eugenics rested on some dubious scientific conclusions, but it held powerful attraction for elites who wanted to believe that modern science could confirm and support their position of privilege and superiority. From the beginning, eugenics was not only a question of scientific hypothesis and evidence, but it also was bound up in a host of assumptions about social and economic status that supported a vision of achievable racial perfection.

Though Galton's statistical work was the basis for small, disconnected eugenics movements that sprouted in various spots around the globe, the movements did not draw significant support until the work of the Austrian monk Gregor Mendel (1822–1884) was rediscovered in 1900. Mendel's 1865 results about the hereditary assortment of dominant and recessive traits in peas stimulated a renewed confidence in the predictability of human heredity. Mendel's theories of inheritance were based on elements that were not directly observable, since a characteristic, such as hair color, that was visible to the human eye might be the genetic result of a combination of both a dominant and a recessive factor where the recessive trait was simply masked by the dominant one. So, while Mendel's work provided a method for predicting heredity, it necessarily included some speculation outside of controlled experimental conditions. More important, however, was the widespread and false assumption that complex human heredity would follow the same single-trait patterns that Mendel's simple pea plants exhibited. However, in the early twentieth century, Mendelian laws of inheritance and August Weissman's hypothesis—that "germ plasma" remained the same from one generation to the next, and thus traits would too—appeared to hold the key to understanding and manipulating the laws of human heredity.[11] Energized by these developments, biologists in the United States went to work trying to distinguish dominant traits from recessive traits in human heredity so that we would all be able to understand—and predict—which traits each mother and father would pass down to their children. The result was an emerging discipline that, in 1905, William Bateson termed "genetics." Much of the work in this new field was done independently of the nascent eugenics movements, but the results would easily, and quickly, be appropriated.

Simultaneously, the period between the turn of the century and World War I was marked by the establishment of a number of institutions that helped to crystallize eugenics as a distinct movement within the United States. The Harvard biologist Charles Benedict Davenport (1866–1944) played a particularly significant role in these developments. By 1904, Davenport had successfully lobbied the Carnegie Institution of Washington to provide funding for the Station for Experimental Evolution at Cold Spring Harbor on Long Island. During his time as director of the research center, Davenport conducted breeding experiments with numerous animal species using Mendelian principles. Shortly after founding Cold Spring Harbor, Davenport began to focus his work on human genetics and launched plans to make the laboratory a center for propaganda about eugenics. In addition to becoming the secretary of the committee on eugenics at the American Breeders' Association, Davenport sought the financial support of philanthropist Mary Harriman, wife of E. H. Harriman, the railroad magnate. Harriman's financial assistance allowed Davenport to purchase a location for the Eugenics Records Office (ERO). Over the next three decades, it would become the preeminent facility in the United States for the promulgation and dissemination of eugenics findings.

Davenport in turn appointed Harry H. Laughlin (1880–1943) as superintendent of the ERO. A school superintendent from Iowa with a keen interest in agricultural breeding experiments, Laughlin eventually earned a doctorate from Princeton in 1917. Under his direction, the ERO trained an army of eugenics researchers, often women, to collect family histories of "deformatives" such as criminals and the mentally ill. Additionally, the ERO willingly accepted submissions of information from individuals and families around the country who wished to document their hereditary background.[12] Perhaps more so than Davenport, Laughlin would prove to be the major figure in public policy initiatives associated with eugenics. During the 1920s, he played a key role in the passage of immigration restrictions, in the legal battle to affirm the constitutionality of involuntary sterilization for institutionalized persons, and in the general effort to place eugenic sterilization statutes on the books of each state and commonwealth in the country.

The Science of Civic Worth

After the initial *AER* exchange on vasectomy, eugenics received periodic attention in more-popular Catholic press venues. This coverage brought together questions about scientific evidence and authority with concerns about racial and class prejudice. While sterilization statutes applied negative eugenic measures to inmates of public institutions, the larger notion of promoting the reproduction of some persons and discouraging the reproduction of others required answering a knottier question—how to differentiate the fit from the unfit. Most eugenics propaganda pointed to the small but growing number of eugenics experts as the appropriate persons to make these distinctions. Fortified with the data and statistics collected at well-known institutions, such as the Eugenic Records Office, and backed by the lustrous veneer of scientific truth, scientists appeared to be able to discern what mental, physical, and behavioral traits marked an individual as a hereditary threat to the population. In an early twentieth century wrapped up in the endless answers that science and technology seemed to provide, it was easy to feel the pull of such thinking.

Some Catholic authors recognized the lure of eugenic thinking, even as they distrusted the shiny veneer of science. They felt that their coreligionists were prime targets for eugenic policies, particularly the many working-class, non-Anglo-Saxon Catholics now spread throughout America. Subsequently, when writers in the more-popular Catholic journals—such as *Catholic World*, published by the Paulists, and *America*, published by the Jesuits—dealt with eugenics, they tended to focus on scientific details. They lamented the ambiguity of eugenic definitions. Their critiques demonstrated skepticism of the language of scientific certitude and an uneasiness about the uses to which science was being put. These writers typically rejected that purely physical or hereditary solutions could solve social problems. But at the same time, Catholic writers did not altogether reject the goal of racial improvement. Rather, they took the opportunity to join the growing tide and articulated a version of positive eugenics that incorporated their values.

The tinge of hostility toward scientific experts evident in such writing is part of a larger story about the ways that the Roman Catholic Church responded to the multiple challenges to its authority during the second half of the nineteenth century. When published in *On the Origin of Species* in 1859, Charles Darwin's theory of evolution by natural selection of random variation caused a good deal of concern and upheaval in Christian circles, including among Catholic scientists and nonscientists alike. Christian commentators adopted a range of ideas from Darwin's work, often citing evolution as the mechanism of God's creation—and just as often reserving the notion of a special creation for human beings while accepting evolution for animals.[13] Reception of Darwin's work in religious circles became the impetus for the growing use of "warfare" as a metaphor to explain the relationship between science and religion, increasingly strained as the nineteenth century came to a close.[14]

A host of research in the history of science has worked to defuse this warfare image, arguing that the actual interaction was far more complex than can be captured in a metaphor of combat or consensus, or even in a single narrative.[15] However, to argue that there was not a war between the

Roman Catholic Church and science is not to say that there was no sense of crisis within the church at the end of the nineteenth century, but rather to say that it was not simply in response to new theories and discoveries in the biological sciences. Vatican authority experienced tremendous political challenges, beginning with the upheavals of the French Revolution and continuing to the revolutions of 1848 and to the efforts to unite Italy into a single state that encroached on the Papal States. These challenges called into question the Vatican's relationship with traditional regimes and promoted widespread anti-clericalism. Also, the rationalism of the Enlightenment led to "higher criticism" of the Bible that no longer treated the scripture as if it were literally true. Combined with discoveries in geology that undermined the veracity of the creation story, these intellectual trends caused the Vatican to react with great distress to the erosion of its traditional base of authority. All of this uncertainty led to an increasing emphasis on centralization of papal authority and control that lasted well into the twentieth century.[16]

In 1878, Pope Leo XIII began a twenty-five-year reign that had profound effects on the way that the institutional church responded to a rapidly modernizing and industrializing world. He took the seat of Peter at a time when ultramontane theology (a conservative movement that championed papal authority), dogmatism, and antirationalism dominated thinking in the Vatican. It was from this perspective that in 1879 Leo issued the *Aeterni Patris,* calling for a return to a scholastic philosophy which predated what he saw as the errors of the Reformation and the Enlightenment.

During the late medieval era, the Scholastics—St. Albert the Great, Roger Bacon, St. Bonaventure, and St. Thomas Aquinas—struggled with the challenges that Greek and Arabic philosophical systems represented for Christianity. Two of the key results of this struggle were a growing appreciation for Aristotelian thought and an effort by many philosophers to incorporate "pagan" truths into mainstream Christian thought. No one was more successful at this synthesis than St. Thomas Aquinas, who saw no need to reject knowledge arrived at through reason, since it properly complemented the truth revealed through faith. Thus, his intellectual synthesis made room for a balance of faith and reason in the attempt to apprehend the complexity of the world, both natural and supernatural. In 1777, Salvatore Roselli, a professor at the College of St. Thomas in Rome, produced a six-volume scholar's edition of Thomas's work, *Summa philosophica,* which greatly influenced the revival of Thomism throughout Europe. Catholic scholars in western Europe drew on Thomas's synthesis to refute the host of thinkers who maintained that reason alone was sufficient to arrive at philosophical truth.[17] Thomas's system of philosophy made a place for both faith and reason, but it was profoundly different from the individualistic, experimental, and questioning stance of modern science.

In *Aeterni Patris* the pope addressed the Catholic hierarchy on the necessity "of taking up the study of philosophy which shall respond most fitly to the excellence of faith, and at the same time be consonant with the dignity of human science."[18] By rendering faith and reason complementary, scholastic philosophy provided a language with which Catholics could systematically refute secular philosophy without appearing to be blindly driven by faith. Leo's concern with rationality and faith extended particularly to the sciences. He maintained that rather than being hostile to the sciences, the Scholastics embraced them, because "they well understood that nothing was of greater use to the philosopher than diligently to search into the mysteries of nature and to be earnest and constant in the study of physical things."[19]

Despite the emphasis on Thomism, a number of Catholic scientists tried to accommodate Darwin's findings in ways that did not conform to the dominant modes of thinking. For the most part, this involved accepting evolution as the method for God's creation while reserving a special place for human creation. While there were lingering questions about the place of humans in the evolutionary scheme, by the 1890s there was quite a good deal of agreement among Catholic scientists that evolution was a viable hypothesis with which they needed to reckon, and one which they could fruitfully address.

Even with this widespread acceptance among Catholic scientists, the reaction in Rome continued to be negative. For Americans, the Vatican's adamant stance was epitomized by the treatment of John Zahm, a physicist at the University of Notre Dame whose *Evolution and Dogma* (1896) was denounced by the Sacred Congregation of the Index, the Vatican office that investigated Catholic publications that were thought to contain errors. Zahm was a prime example of the intellectual movement that explored new approaches to a range of questions, including evolution and historical criticism, but that most importantly moved away from the neo-scholastic approach that came to dominate Catholic intellectual life during Leo XIII's papacy. After minimal investigation, the Congregation issued a private condemnation to Zahm rather than a public rebuke. Zahm submitted to Roman authority, withdrew the book from publication, and refrained from speaking further about matters of faith and science.[20]

This episode in American Catholic history was part of a larger modernist controversy that had far-reaching implications for the ways that Catholic thinkers approached the intellectual and social questions of the day. Some American Catholic priests—including those who worked as scientists and theologians—purposefully tried to adapt their religious ideas to the great arc of modernist thought during the last decade of the nineteenth century and the opening decade of the twentieth.[21] Zahm's experience with the Vatican foreshadowed an increasingly intolerant response to theological innovation. In 1907, however, Pius X closed the door on these novel approaches by condemning theological modernism. In *Pascendi Dominici Gregis* he reacted against the limited intellectual forays of recent years and declared that "those who favour Modernism either by extolling the Modernists or excusing their culpable conduct, by criticising scholasticism, the Holy Father, or by refusing obedience to ecclesiastical authority in any of its depositaries; and towards those who show a love of novelty in history, archaeology, biblical exegesis, and finally towards those who neglect the sacred sciences or appear to prefer to them the profane," should be excluded from leading Catholic universities or seminaries, from serving on their faculties, or from being considered for Holy Orders.[22] The result was a quieting of an already limited movement. In a larger sense, the condemnation also resulted in a tendency among American Catholics to regard intellectual systems other than scholasticism—not just those concerned with theology, but also those that prized scientific knowledge over other ways of knowing—with suspicion.

Thus, after the initially narrow discussion of the morality of vasectomy in the *AER* in 1910–11, Catholic writers began to question the larger claims and goals of the eugenics movement. They distrusted eugenic principles and definitions because they were dubious of what passed for scientific method in eugenics circles and thus doubted the validity of eugenics research. Writing in 1912, for example, Thomas J. Gerrard questioned the ability of eugenicists to articulate the specific ends of their movement. His article, which appeared in *Catholic World*, pressed eugenicists to clarify what Francis Galton had in mind when he held up "civic worth" as the end to be obtained by positive eugenic philosophies. Determining what constituted "civic worth" would involve making clear distinctions between those persons who contributed to the social welfare and those persons who detracted from it. The trouble was, Gerrard could not find a definitive list of traits or social movements that contributed to the "civic worth." Consulting the work of the prominent British eugenicist, Dr. C. W. Saleeby, Gerrard was still befuddled. Unable to quite grasp the concrete application of eugenics in the world, he submitted his own tongue-in-cheek summary of positive eugenics philosophy: "It matters not what you choose to strive for, provided only that you use the modern eugenic principles to help you in your striving."[23] Gerrard was deeply troubled by the vagueness of the stated ends and aims of positive eugenics, in part because he suspected that they were a guise for the solidification of a discriminatory social hierarchy.

Others were troubled by the use of Alfred Binet's intelligence test and the methods employed by Dr. Henry H. Goddard, the director of research at the Training School for Feebleminded Boys and Girls, a prominent site of eugenics investigations in Vineland, New Jersey. In addition to coining the term "moron," Goddard used Binet's intelligence scale to popularize the notion that intelligence was passed from generation to generation as a single trait that followed Mendelian laws of inheritance. In his 1912 account of the Kallikak family, Goddard and his assistants traced two family lines, one "good" and one "bad," to a single father. They attributed the hard luck of the "degenerate" family to an ill-considered mating of the progenitor, Martin Kallikak, with a supposedly feeble-minded tavern maid. His later union with an upstanding Quaker woman fared much better.[24] Despite what stand out to twenty-first-century readers as blatantly subjective judgments in his studies, the elegant simplicity of Goddard's conclusions were a great boon to the eugenics movement. In short, they made it possible to read the future: as long as you knew the mother and the father, you would know the intelligence of any child produced by their union. Significantly, Goddard's formulations suggested that degenerate traits were dominant, and extremely dangerous for the well-being of future generations.

In 1916, the Reverend Thomas Verner Moore (1877–1969), a Catholic psychiatrist, used his address to the annual National Conference of Catholic Charities meeting to question Goddard's methodology in the Kallikak study. Criticizing the assumption by Goddard and his fieldworkers that the mental abilities of their subjects was easily discernible through general observation and questioning, Moore explained to his audience, "It seems to me Dr. Goddard and his social workers must have been very often deceived, and so they have been able to fit the facts to a preconceived hereditary scheme. . . . They did not rely upon objective standards but subjective impressions, and their very earnestness has led them astray."[25] Moore believed that attributing feeble-mindedness to hereditary factors restricted the courses of treatment available, resulting in all hope of any effective treatment being abandoned. His work with mentally disabled children emphasized guidance, clinical diagnosis, and a willingness to apply modern methods of medicine and psychology within the context of traditional Catholic charitable and welfare institutions.[26]

As part of the Progressive reform movements of the era, welfare workers were deeply influenced by an increasingly scientific approach to social problems, one that coincided with the emergence of a set of academic disciplines which sought answers outside of the particularities of history. Practitioners in these new areas of inquiry—political science, economics, sociology—looked for natural laws to guide their analysis and to develop systematic modes of explaining human behavior. Such methods purported to result in the discovery of objective truths about the social world.[27] In addition to the establishment of the social sciences, notions of scientific management and professionalization also changed the work of charity.

These influences were apparent quite quickly within Catholic social work, particularly with the establishment of the National Conference of Catholic Charities in 1910. Once the NCCC was formed, the leadership began to push for reform and a standardization of practices, which would only culminate after World War I, when in 1920 John Burke, of the National Catholic Welfare Council and the National Council of Catholic Women, moved to form the National Catholic School of Social Service as a part of the Catholic University of America. Other Catholic schools also moved to create programs to train Catholic social workers in the professional practice of case work.[28] The efforts of Moore and other Catholic thinkers were key to balancing these new professional principles and the enduring values of Catholicism.

By challenging Goddard's findings, Moore struck a blow at one of the major scientific tenets of the early eugenics movement—that unilinear heritability of mental capacity was the cornerstone rationale for shaping social development through controlling reproduction. If a single gene did not control intelligence, then eugenicists had no hope of accurately predicting the future mental quality of offspring. By implication, if intelligence was not solely a question of heredity, the sterilization movement was purely guesswork. Admitting the possibility that mental deficiency could frequently be traced to non-congenital factors also necessitated admitting the possibility that therapeutic and preventative measures would be more humane and effective than forced sterilization. Moore's arguments on these fronts threatened the entire scientific foundation of the eugenic approach to mental deficiency. Unfortunately, Moore's work had a somewhat limited audience, and his critique failed to garner much of a hearing outside of primarily Catholic circles. Nonetheless, when Catholics began to lobby against eugenic policy initiatives, they turned to Moore's work to bolster their arguments.

In 1917, in a strongly worded article for *Catholic World*, Henry Somerville echoed Moore's arguments by denying that eugenicists had actually constructed a reliable definition and test for hereditary deficiency.[29] Somerville objected to the definition of feeble-mindedness, as articulated by Goddard, as "a state of mental defect existing from birth or from an early age and due to incomplete or abnormal development, in consequence of which the person affected is incapable of performing his duties as a member of society in the position of life to which he is born."[30] Somerville read this definition as an indication of the arbitrary nature of feeble-minded diagnoses and the tendency of eugenicists to deny the importance of social environment in their judgments. Thus, he rejected the materialistic perspective in eugenics work by turning to Catholic teaching. "The sacredness of the individual" he explained, "does not depend upon his economic value to society, and the individual does not cease to have human rights because he suffers from mental or physical defects. Among the most fundamental of human rights is the right to marriage, and this carries with it, in the Catholic if not the eugenist view, the right to parenthood."[31] In light of Catholic teaching, Somerville urged his readers to resist the calls for compulsory sterilization and segregation of the so-called feeble-minded as a grave violation of human rights based on ambiguous and unscientific theories. Somerville's emphasis on the economic motives behind eugenic theories came up again and again as Catholics reflected on the goals of the movement.

Considering these topics led Catholic writers to develop their own version of positive eugenics that relied upon Catholic teaching to encourage the development of a morally sound population and as an alternative to hereditary principles. In his 1912 article, Thomas J. Gerrard went beyond criticisms of Francis Galton and proposed that the Catholic method for determining proper social behavior—"right reason duly informed by Divine Will"—provided a more reliable means to improve the race than eugenics principles. Given St. Paul's teaching that Christians are the "adopted sons of God," Gerrard argued that they were free to exercise their powers of reason under the guidance of divine grace and the Holy Spirit. Moved by the Holy Spirit, Christians were supposed to exhibit both the theological virtues (faith, hope, and charity) and the cardinal virtues (prudence, justice, fortitude, and temperance) in their social life and behavior, which "makes for an infinitely higher responsibility and dignity than is possible in any naturalist system." Gerrard explained that "from the theological virtues . . . there flow directly those habits of mind and heart by which every sound eugenic reform is brought about, and by which every sound principle of positive eugenics is applied to the race for its higher well-being."[32] Here we see the preeminence of individual reason and decision making in Gerrard's understanding of the path to improvement for a community. For Gerrard, the principles of Christian morality would lead to a moral regeneration of society that would achieve the ill-articulated goals of eugenicists without resorting to invasive measures. As his writing makes clear, the goal of racial improvement was a sound one, but the grounds for making judgments about human worth, and the means for bring about that racial improvement, lay at the heart of the conflict between Catholics and eugenicists.

Thus, the ambiguities in eugenics rhetoric and definitions allowed Catholics to attempt to forge their own version of "positive eugenics." As members of a marginal group within the dominant culture, some Catholic writers felt that they had to respond to popular eugenics rhetoric by showing the compatibility of Catholic teaching with the seemingly benign eugenics aim of creating a strong healthy population—a population that they did not clearly characterize as being derived from any particular heritage. Fundamentally, however, this task required a rejection of traditional eugenics standards and means for effecting change. To create a Catholic positive eugenics, authors would have to provide an alternative process for determining and encouraging activity that would improve the social welfare. Rather than pushing social reform based on hereditary principles, Catholic thinkers argued that Catholic doctrine provided a reliable guide for social reform. Focusing on principles of free will and the influence of the environment in bringing about social change, Catholics suggested that a sound moral code could do more to improve the American population than any process of selective breeding.

In claiming this common ground with eugenicists on racial improvement, these authors invite us to speculate on their meaning in using the term "race." Working in a period of great upheaval in the American racial landscape, these Catholic writers left themselves open to being interpreted as sharing the racial assumptions of eugenicists, who valued Anglo-Saxon and Nordic racial heritage above all others. However, Gerrard uses the terms "human," "man," and "race" almost interchangeably. Also, the makeup of the contemporary Catholic community suggests that Catholic writers would be unlikely to support claims for an Anglo- and Nordic-dominated racial hierarchy. Between the 1880s and the 1920s, the American Catholic community absorbed and adjusted to the dramatic influx of immigrants from eastern and southern Europe by establishing ethnic parishes and an expanding social service apparatus. That the 1911 federal Dillingham Commission's report on immigration noted forty-five races of people coming to the nation, thirty-six of whom were from Europe, would not have surprised any Catholic residents of the urban northern part of the United States, who bore witness daily to the fine distinctions among immigrant racial groups such as Celts, Italians, Slovaks, Poles, and others. Given the small numbers of African American Catholics, this world of racial variety would appear mostly white to contemporary eyes. Assumptions about these racial distinctions continued for decades and created a complicated racial terrain. Inevitably, the growth in racial diversity resulted in tensions, and highlighted racist assumptions on the part of parish priests and their predominantly Irish and German bishops.[33] But at the same time, a still strong climate of distrust and anti-Catholicism among native-born, Anglo-Protestants made Catholics wary of prejudice from without.[34]

Birth Control

On guard against immorality and reform movements that unfairly discriminated against their co-religionists, Catholic writers, scientists, and social scientists raised social objections about the movement to provide unfettered access to literature about contraception and birth control devices. Those objections were similar to the ones they raised about eugenics. As efforts to purposefully control reproduction, the eugenics and birth control movements often appeared linked to Catholic thinkers in their efforts to solve social problems through revising reproductive practices.

Both movements were invigorated by the reform spirit of the Progressive era, whose optimism animated efforts to improve huge swaths of American society. For some this meant far-reaching legislative schemes to transform environmental conditions, and for others it meant fostering widespread changes in personal behavior. Hence, the targets of the Progressive spirit were as varied as the era's many reforms. Some middle-class activists sought reform in order to rescue the political system from corruption. Others worked diligently to improve the working and living conditions of the working classes in urban America. Upton Sinclair and Lewis Hine focused attention on industrial working conditions, while Jane Addams and Ellen Starr Gates focused on the needs of the urban poor and new immigrants. Some activists, such as Big Bill Haywood and Elizabeth Gurley Flynn, who championed socialism and labor organizing, took a more radical path to reform. The modern birth control movement grew up in the soil of this more-radical path when Margaret Sanger set out to challenge the public silence about contraception in the first decade of the twentieth century.

Despite its shocking appearance on the public scene in America just before World War I, a more general conversation about controlling reproduction can be traced back at least a century and a half prior, to a distinct decline in the birth rate of eighteenth-century France. There, a full range of intellectual, ideological, and religious shifts allowed for the discussion and eventual acceptance of reproductive limitation. By the close of the same century in England, Thomas Malthus made a significant intellectual impact with his thesis about the dangers of overpopulation. While Malthus did not suggest contraception as a solution to the problem he observed, others—such as Jeremy Bentham, James Mill, and Francis Place—eventually did. In the United States, the call for controlling reproduction was taken up by Robert Dale Owen and Charles Knowlton in the 1830s. At the same time, the invention of vulcanized rubber by Charles Goodyear, in 1837, vastly increased the variety and effectiveness of contraceptive devices. These devices—condoms, pessaries, sponges, etc.—were widely advertised in the American press until the passage in the 1870s of the Comstock laws, obscenity statutes championed by former postal inspector Anthony Comstock.[35]

Comstock's influential reticence, and his desire to enforce a public sense of propriety and an environment of silence on sensitive topics like contraception, was shared by Catholic clergy in the late nineteenth and early twentieth centuries. Parish priests across the United States were reluctant to pursue the question of contraception, both in the confessional or from the pulpit. Counseled by manuals to be cautious about probing parishioners about sexual sin, pastors mostly refrained from inquiring about marital contraception altogether, and never preached about it.[36] This climate of general reticence began to change when Margaret Sanger's efforts made contraception a public issue.

The charismatic and intellectual force behind the birth control movement in the United States, Sanger (1879–1966) made herself the heart of this public conversation for over five decades. As a visiting nurse and midwife in New York, Sanger was intimately familiar with the emotional and physical distress that frequent pregnancy might bring for women. She was simultaneously radicalized by her work with the Socialist Party, imbibing socialist and labor politics through associations with the Industrial Workers of the World, the Lawrence textile strike, and (through her relationship with Emma Goldman) feminist ideology. In 1912, Sanger began writing a column for *The Call*, the New York Socialist daily, which dealt with sex education and women's health. From the column "What Every Girl Should Know," Sanger moved on in 1914 to write and publish *The Woman Rebel*, a magazine designed to dislodge the prevailing public silence on sex education and contraception. The result was swift: after the first issue was published and distributed, she was arrested for violating the Comstock laws. Rather than prepare for her trial, Sanger penned the pamphlet *Family Limitation*, in August 1914, and then fled to Europe for ten months. While Sanger avoided prosecution in Europe, she gathered information about the latest and most effective contraceptive methods there, and her supporters circulated copies of *Family Limitation*.[37]

Margaret Sanger's birth control philosophies enabled women to envision lives for themselves that were not inextricably tied to motherhood. With adequate birth control, women could acknowledge their sexual desires without being trapped in an endless cycle of pregnancy and childbirth. They would be free to create meaningful roles for themselves based on human equality, not on their identity as mothers. With that goal—nearly unfathomable for so many early-twentieth-century Americans—in mind, Sanger packed *Family Limitation* full of frank descriptions of various contraceptive devices, including illustrations and instructions as to their proper usage. In the pamphlet, Sanger suggested that wage workers neither really desired, nor could they support, more than two children, exclaiming: "It is only the workers who are ignorant of the knowledge of how to prevent bringing children in the world to fill jails and hospitals, factories and mills, insane asylums and premature graves."[38] She wanted to provide the working classes with information about safe and effective contraception—information that had been available to the middle and upper classes for years—which would allow them to limit the size of their families, therefore freeing them from undue economic hardship and, in particular, freeing women from the debilitating trauma of perpetual pregnancies.

Sanger's campaign was shocking enough to shake the Catholic clergy from its reluctance to speak out on the matter. One prominent Catholic who stepped forward was the Reverend John A. Ryan (1869–1945), a progressive priest and political economist. While teaching moral theology at the Catholic University of America in 1916, Ryan published an article, also entitled "Family Limitation," in the *AER*. The article set out the objections of the church to artificial means for limiting the number of births and provided the clergy with the justifications for those objections. Ryan laid out the main natural law argument against artificial contraception: "The generative faculty has as its specific and essential end the procreation of offspring. That is the object which explains and rationalizes this particular faculty. When the faculty is so used that the very use of it renders the fulfillment of its very purpose impossible, it is perverted, used unnaturally, and therefore sinfully."[39] He acknowledged that immediate lived experience might make arguments for birth control based on the promotion of social welfare or race betterment more appealing than the rationale of natural law against contraception, but he argued that natural law nevertheless provided the correct answer on this moral issue.

In an effort to speak to those more immediate and less theoretical concerns, Ryan outlined what he saw as some of the practical consequences of birth control. Ryan was convinced that contraceptive practices would lead to a diminishing population, a phenomenon he identified as "race suicide." First used by the sociologist E. A. Ross in a 1901 essay called "The Causes of Race Superiority," and then championed most memorably by Theodore Roosevelt, the term "race suicide" encompassed two strains of thinking. On the one hand, individuals who sounded the alarm were concerned about the high birthrates of undesirable racial groups, namely, new immigrants; almost immediately upon the upsurge in immigration in the 1880s, a fear grew that these newcomers could and would outbreed native-born Anglo-Americans. The corollary of this argument was that native-born, middle-class women were neglecting their duty to produce a sufficient number of offspring. In the minds of those who warned of race suicide, a massive surrender to luxury and indulgence on the part of the middle class was literally destroying the prospects for vigorous future generations. Only a retreat from luxury would save the race; contraception was the epitome of such dangers.[40]

Unlike many eugenicists who understood ethnic groups as distinct racial groups with varying biological qualities and worth, Ryan, like Gerrard, appears to have used the word "race" to refer to the human population as a whole. Invoking the widespread fear about falling birth rates, Ryan turned eugenicists' racialist tendencies against them. He tacitly acknowledged the express fears of xenophobes who worried that new immigrant birthrates would swamp the native-born middle classes. Arguing that the working classes might "function as the 'saving remnant' of civilization" if they could resist the temptations of luxury that birth control represented, Ryan resisted the deep prejudice that pervaded nativist arguments about immigration, differential birth rates, and biologically determined ability. Rather, Ryan's use of the term "race" was inclusive of diversity and recognized the social and cultural tensions that accompanied what Matthew Frye Jacobsen has called "variegated whiteness" during this period.[41]

Seemingly uninterested in the project of further defining racial hierarchies, Ryan felt that material comfort propelled the birth control movement, an inclination that would inevitably do irreversible harm to the moral constitution of the population. Ryan maintained that since the birth control movement was underwritten by the principles of "dislike of sacrifice and disinclination to painful effort" and "the indefinite increase and variation of pleasant physical sensations," the movement created an "atmosphere of ease, egotism, materialism; which is generally fatal to the development of those moral qualities which are essential to high mental discipline, disinterested service of the neighbor, self-denying application, and the sustained pursuit of any great and beneficent ideal."[42] Unlike Margaret Sanger, who saw access to birth control as a key facilitator of women's independence, in Ryan's opinion, the dissolution of the self-sacrificing character was an inevitable effect of the spread of birth control practices. In fact, he already saw evidence of such dissolution in the behavior of the members of the upper and middle classes, who made use of contraceptive devices with much more frequency than the working-class population.[43] Ryan's comments also responded to a resurgence of Malthusian thinking among birth control advocates who argued that the population was growing far beyond its ability to provide for itself. Finding evidence of large families among recent immigrants and working-class persons, birth control advocates attributed the hardships of their social position to their reproductive habits, in turn blaming poverty on the indigent. Ryan, in turn, spoke out for a course of social reform centered on a just distribution of resources that would allow all individuals to provide for the basic material needs of their families.[44] Thus, Ryan viewed the birth control movement as a challenge to the complex web of needs and obligations that connected the family to the larger social world. Ryan spent his entire career advocating for a living wage for working families, and he saw this as an alternative to calling for birth control as a way to ease the material burdens of the working classes.[45]

In focusing attention on what each family needs—and what we as a society need to give them—rather than on seeing those needs as a burden to society, Ryan rejected the tendency of eugenicists to focus on an individual's ability to contribute productively to society. While addressing the church's stance on birth control, Ryan made an explicit commentary on a fundamental difference in the attitude of eugenicists and Catholics toward intrinsic human worth and the "welfare of the race." Contrary to proponents of eugenics, Ryan explained that "the Church always looks upon the spiritual and moral side of individuals and institutions as much more important than their physical aspects or consequences. . . . Consequently her viewpoint is infinitely removed from that of those practical atheists who measure the worth of a subnormal person by the same standard that they apply to a dog or a horse." He questioned the ambiguities of eugenic claims that "subnormal" persons endangered the social welfare. As a result, Ryan suggested that the goal of preserving the "welfare of the race" was "a mere abstraction that corresponds to no definite idea; or it means the welfare of the fortunate majority who do not desire the inconvenience of helping to support any considerable number of defectives." In Ryan's thinking, calls for contraceptives and sterilization for the "unfit" or "feebleminded" were simply a guise for self-interest and material opportunism.[46] Where birth control advocates suggested that access to contraceptive devices would ease the burdens of the working classes, resulting in an improved standard of living, Ryan saw a shortcut to economic relief that relied upon a willingness to transgress fundamental laws of nature rather than demanding that employers adequately provide for their workers. Where eugenicists saw an answer to the growing population of degenerates that they saw as a drain on resources, Ryan saw an unwillingness to respect the intrinsic value of every human being, regardless of his or her productivity.

The War and After

On April 2, 1917, the United States officially entered the Great War. Across America, domestic mobilization consumed the energies of many elements in public life; eugenics faded into the background. Over the next eighteen months, state governments dealt with little in the way of eugenics legislation, with only Oregon and South Dakota approving sterilization statutes.[47] For ordinary Catholics, still maligned in some quarters as harboring loyalties that superseded their allegiance to the nation, the war represented an opportunity to prove that they were patriotic Americans. The church's bishops in the United States also realized the opportunity, throwing their support behind the mobilization efforts and forming the National Catholic War Council to attend to the needs of Catholic troops and to support Catholic war work. Showing a striking degree of unity, the members of the Catholic hierarchy supported the war through the council with fundraising drives and an effort to create a range of educational materials for new immigrants that introduced them to American democratic values.[48]

As the war approached its close, the Administrative Committee of the War Council began to consider the challenges of postwar life for the United States. Concerns about peacetime readjustment abounded, especially about how returning soldiers would make their way in an unstable economy. The Reverend John O'Grady, secretary of the Committee on Reconstruction, turned to John Ryan to draft a statement that would take into concern economic questions from a Catholic perspective. The result was the "Bishop's Program on Social Reconstruction," released in early 1919.[49]

Released by the War Council, Ryan's statement in effect became the official position of the American Catholic Church, emphasizing its commitment to the notion that "the laborer is a human being, not merely an instrument of production; and that the laborer's right to a decent livelihood is the first moral charge upon industry."[50] As such, it called for a commitment to the principle of a living wage, an endorsement of the worker's right to organize and bargain collectively, a call for some guarantee of livelihood for returning soldiers, a commitment to antitrust and public competition to fight the monopolies of private industry, an endorsement of widespread ownership of stock and participation in management by workers, access to vocational training for those who needed it, and a guarantee of a minimum wage as a matter of justice. Furthermore, the statement strongly endorsed bringing an end to child labor. And although calling for women to relinquish jobs in industry to returning soldiers, the plan urged a just wage for women who needed to remain in the workforce. By releasing this statement, the War Council endorsed Ryan's broader vision of social justice through economic justice.

Later in 1919 the US Catholic hierarchy held a general meeting attended by the heads of each American diocese, a gathering the likes of which had not occurred since the Third Plenary Council in 1884. The September meeting resulted in a Pastoral Letter outlining the hierarchy's vision for the church in the United States after the war. The letter received a wide hearing, with portions of it read from the pulpit of every parish in the United States on George Washington's birthday in 1920.[51] Reviewing the progress of the church in spiritual and temporal matters in the preceding thirty-five years, and covering such issues as the growth in Catholic education, Catholic societies, and the Catholic press, as well as setting out the bishops' thinking on a wide range of secular issues, the letter touched on a number of issues connected to the eugenics movement. For example, the pastoral addressed the proper role and authority of the state, claiming: "The end for which the State exists and for which authority is given it, determines the limit of its powers. It must respect and protect the divinely established rights of the individual and of the family."[52] This concern for the centrality of the family, and the need for its protection, was echoed in a specific rejection of family limitation. The bishops explicitly linked birth control to eugenic aims, explaining: "The selfishness which leads to race suicide with or without the pretext of bettering the species, is, in God's sight, 'a detestable thing.' It is a crime of individuals for which, eventually, the nation must suffer. The harm which it does cannot be repaired by social service, nor offset by pretending economic or domestic advantage."[53] Together, these statements on the proper role of the state and its duty to protect the nuclear family, and on the duty of those families to welcome children, set the tone for much of the subsequent Catholic response to eugenics policy initiatives. Also, the text's reference to "bettering the species," rather than "the race," provides further confirmation that the hierarchy rejected the racialist assumptions embodied by the eugenics movement.

The birth of the NCWC inaugurated the next stage in the evolution of the Catholic Church in America, and in turn would play an essential part in the church's relationship with the eugenics movement. As we shall see, the momentum of the movement would in the next decade eclipse its influence prior to World War I and eugenics advocates would have a hand in policy decisions in a majority of states. In retrospect, the pre–World War I eugenics movement had not yet reached the apex of its power, as it would in the late 1920s. And the Catholic Church's engagement with that movement was, in turn, still in a critical stage of formation and debate.

Crucial in these prewar years, however, was the range of disagreement that Catholic writers and thinkers offered, and equally important were the realms in which they found accord with eugenics thinking. For here a pattern is laid down that will prove to be persistent over the course of decades of Catholic engagement with the eugenics movement. The popular rhetoric about improvement made eugenics language hard to resist. Hence, some Catholic thinkers tried to adapt the terms of the movement to their own perspectives on what facilitated social improvement. For the most part, this meant an emphasis on moral decision making, rather than biological determinism, as the path to a transformed community. This rejection was anchored in a willingness to question the answers that science appeared to offer about human worth. Faced with the emergence of the birth control movement, pronatalism received an increasing emphasis from Catholic writers and members of the hierarchy, offering a point of common ground with eugenicists. However, here and in the future, these voices refused to make the hierarchical racial distinctions that were at the core of eugenics thinking. Instead, they pointed to the need for economic justice—an environmental solution—in the face of the needs of the working classes, rather than reproductive restriction. From the base of these initial interactions, we can see elements of agreement and conflict that would remain in play in Catholic perspectives on eugenics policies over the course of the next thirty years.

TWO

Cooperative Clergy? Catholics in the American Eugenics Society

In 1921, among the wildlife and the dioramas that filled the galleries of New York's American Museum of Natural History, the noted Yale economist Irving Fisher joined with Charles Davenport, Madison Grant, C. C. Little, and Harry Olson at the Second International Congress of Eugenics to plan a national organization to promote eugenics in the United States. Taking their cue from the success of national eugenics organizations in Norway, Sweden, and England, the American participants began to appreciate the advantages of a single organization that could coordinate eugenics efforts across the country. The postwar meeting marked the first steps in the creation of what would eventually become the American Eugenics Society.[1]

In the summer of 1922, a host of prominent American scientists, educators, religious leaders, and politicians received invitations to become members of the advisory council of the new organization. The words of the recruitment letter clearly expressed the urgency of the group's founders: "The time is ripe for a strong public movement to stem the tide of threatening racial degeneracy." The nation needed to guard carefully against "indiscriminate immigration, criminal degenerates," and those who would participate in "race suicide."[2] While most invitations went to Protestants, a small number went to non-Protestants. New York Democratic and Catholic politician Alfred E. Smith failed to respond to his June 1922 invitation to join the council. However, many individuals did respond, including a number of members of the clergy. The Reverend John Montgomery Cooper (1881–1949), a prominent anthropologist from the Catholic University of America, was the lone Catholic who accepted an invitation.[3] By February 1923, the advisory council was composed of ninety-nine men; each had some sort of interest in eugenics ideology, and many fit the desired eugenicist's profile—native born, middle class, professional.[4]

Though the organization went through a number of name changes, eventually it was incorporated as the American Eugenics Society in 1926. The AES thrived in the charged nativist environment of the 1920s. The group's leadership developed strategies to address their opponents and to convert those who had not yet decided to endorse eugenics as a comprehensive vehicle for national development. They engaged in a process of coalition building that reinforced their sociopolitical salience and ensured that eugenics philosophies would influence a wide range of public policy decisions throughout the decade. The result was a flurry of successful legislative activity in the 1920s.[5]

In the August 1923 issue of *Eugenical News*, the AES published a comprehensive organizational agenda that served as a roadmap for the decade of activity to come in the areas of research, education, and law.[6] The committee's vision for this research was far-reaching: "Research is needed to answer the question whether religion, philanthropy, modern sanitation, and medical progress are really eugenic. If not, how can their vast *social* value be preserved and a *eugenic* value be added?"[7] But this research would only be useful if it could be communicated to the public. With respect to the goals of the educational plan, the authors claimed, "We should endeavor to show that eugenics supplies the most effective and permanent solution to the problems which have been so ineffectively dealt with hitherto by physicians, public health officers, social workers, clergymen, and reformers—the problems of combating disease, disability, defectiveness, degeneracy, delinquency, vice and crime."[8] Thus, the program committee also proposed a full slate of popular education initiatives, including plans to enlist the YMCA, the YWCA, public libraries, the Boy Scouts, and the Army and Navy in spreading the word, as well as holding summer schools and hosting a series of lectures.[9]

In addition to educating the public, the program committee proposed that the American Eugenics Society should work to systematize and develop a "eugenic code" of law, not unlike the sanitary code or the criminal code. The eugenic code would address "laws having for their purpose the conservation of the best stocks of the race." First, the code would provide a host of positive eugenic measures for individual families by instituting laws related to marriage and divorce, illegitimacy, and mothers' pensions, as well as laws encouraging parenthood among those with the best hereditary traits, laws encouraging differential salaries based on family size, and laws concerning the registration of family pedigrees. Conversely, the code would limit the number of defectives by providing laws for their segregation and/or sterilization, as well as laws regulating immigration and amending the existing immigration statutes to add mental and physical requirements to the current literacy tests, plus the establishment of state bureaus of eugenics. Also, the AES proposed to provide for eugenics education, mental testing for children, and the elimination of laws that restricted the distribution of contraceptive information. Finally, the organization took up other policy issues by encouraging laws that would improve international relations and reduce the occasion for war, as well as encouraging reforms on laws dealing with taxation, wages, and labor conditions. In sum, the proposed legal code would offer a comprehensive system of law; it would result in a total reform of American society under the direction of eugenics principles.[10]

The agenda outlined by the AES in the early months of its organization constituted a comprehensive, progressive, perfectionist plan for racial improvement. By regulating the population through the sealing of national borders, and by proscribing the reproductive activities of the current citizenry, the AES proposed to actively shape the future of the nation. The vast ambitions of the AES were matched, however, by a canny sense of strategy. They knew that such a vast restructuring of American life would only be possible through recruiting and converting the members of the so-called superior classes, who held an inordinate amount of social and political power, to a eugenic understanding of the hierarchical valuing of human lives.

Part of the effort to win converts to their agenda, and to soothe the fears of those opposed to eugenics, involved the establishment of the Committee on Cooperation with Clergymen (CCC). The group contained two Catholic members, the Reverend John A. Ryan and the Reverend John Montgomery Cooper. Though eugenicists experienced growing resistance from Catholics on the local and national levels to their efforts to pass and implement sterilization statutes during the 1920s, as we shall see, representatives of the movement specifically and directly engaged Ryan, Cooper, and others in conversation about the eugenics agenda through the CCC and the eugenics press. These conversations vividly illustrate the positions of Ryan and Cooper on a cluster of issues associated with eugenics, including immigration, birth control, feminism, and sterilization. The interactions reveal the degree to which the Catholics involved with the movement embraced the goals of positive eugenics, promoting the marriage and reproduction of superior types through popular education and state-sponsored welfare programs. Though they disagreed with eugenicists' methods of distinguishing between superior and inferior persons, they wholeheartedly endorsed measures that echoed the church's teaching on the centrality of the stable nuclear family and the importance of virtue and sound morality to the survival of that unit.

While the Catholic members of the CCC tended to concur with the pronatalist elements of positive eugenics, as discussed earlier, they parted company with the movement on negative eugenics policies, such as the dissemination of birth control for the poor, immigration restriction based on national origin, and sterilization of the "unfit," that forcibly discouraged the reproduction of certain groups. The gap between these two positions revealed not just a difference of opinion but a different understanding of the meaning and uses of scientific data. Eugenicists generally viewed positivist scientific data and theories as value-free and objective. As a consequence, Catholic moral and theological arguments about reproductive politics were often unconvincing to non-Catholics because they appeared to be subjective and arbitrary, attacking conclusions that eugenicists took for granted as supported by empirical data. Ryan and Cooper frequently engaged eugenicists on the terrain of science itself, questioning the theories, methods, and conclusions of the movement, all of which relied heavily on biological determinism. In seeking to challenge eugenicists on their science, these Catholics engaged in a process that Sander Gilman and Nancy Leys Stepan have called "recontextualization." One of a number of rhetorical strategies by which minority targets of scientific racism resisted that discourse, "recontextualization" involves two processes: "First, the tools of science were used either to prove that the supposed factual data upon which the stereotypes of racial inferiority were based were wrong, or to generate new 'facts' on which different claims could be made. Second, scientific reasoning was used to question the *explanation* of the facts."[11] Challenging eugenicists on their theories and methods, Catholics concerned about eugenics continually questioned the data, and the conclusions that eugenicists drew from that data. More often than not, they argued that the data from eugenics studies were inconclusive and that eugenicists had failed to adequately consider environmental factors in their analysis of social conditions.

The results of the interaction between eugenicists and Catholics involved with the movement—which appear mostly in the pages of the two monthly AES publications, *Eugenical News* and *Eugenics*—suggest that those Catholics mostly rejected negative eugenics policy initiatives and that eugenicists found Catholic moral reasoning unconvincing. These interactions centered on contested notions of scientific methodology and conclusions. Though the leadership of the AES was not wrong to hope that they could find common ground with Catholics in their selective pronatalist stance, each interaction reveals aspects of the contours of an unbridgeable gap between Catholics and supporters of the AES's negative policy agenda. Thus, in the period between 1923 and 1930, we shall see a growing antipathy between eugenics activists and even those Catholics who were the most sympathetic to the movement.

Cultivating a Cooperative Clergy

In the process of building a coalition of support for specific policy measures, the leadership at the AES quickly recognized the importance of reaching out to multiple components of American society. To facilitate the full acceptance and implementation of a eugenics policy and social agenda, they needed the support not only of biologists, anthropologists, and politicians, but also that of persons who worked in social services and healthcare. Most of the elements of the AES agenda called for lobbying on the local level for legislative measures. If such measures had any chance of passing city councils and state legislatures, they would need the support of churches and civic groups. In October 1923, barely six months after its creation, the AES advisory council voted to establish three new committees: the Committee on Cooperation with Physicians, the Committee on Cooperation with Social Workers, and the Committee on Cooperation with Clergymen.[12]

Despite this early recognition of a need for dialogue with the religious community, the CCC did not fully take shape until 1925. The Reverend Henry S. Huntington, brother of the prominent eugenicist and geologist Ellsworth Huntington, was named chairman of the committee, and plans were laid for a small executive committee and a larger general committee of forty persons. The executive committee would be responsible for generating initial plans and then submitting them to the larger committee for feedback. Included among the prominent Protestant members of the general committee were the editor of the *Christian Century* and two presidents of the Federal Council of Churches of Christ in America. Seven of the Protestant members had just been named *Christian Century* "American Pulpit Leaders." While the bulk of the original forty members were Protestant luminaries, the committee also included two Reformed rabbis, Louis Mann and David de Sola Pool, and two Catholic priests, John A. Ryan and John Montgomery Cooper.[13]

The members of the AES executive committee could not ignore the significant place that Roman Catholics held in American society by the middle of the 1920s. In addition to dominating urban politics in a number of significant cities, the sheer numbers of Catholics in the population made them a force to contend with. In 1920, the US Census reported that the national population stood at just over 105 million people. By 1930, that population had ballooned to over 122 million. In the midst of that dramatic increase, the 1926 Census of Religious Bodies found that Catholics constituted 18.6 million people within the national population, with 9.28 million of them concentrated in Massachusetts, New York, New Jersey, Pennsylvania, and Illinois.[14] While this does not seem like an enormous number in comparison with the total population, this concentration of Catholics was lodged firmly in the areas of the country where the most prominent eugenicists lived. Though many eugenicists were good at separating themselves from the messy world around them, especially through their work at major private museums, colleges, research institutes, and universities, the unavoidable reality was that eugenicists were awash in a sea of great ethnic, economic, and religious diversity, of which Catholics were a major part.[15]

The AES must have had some inkling that it had its work cut out for it if it wanted to attract Catholics to the movement since eugenicists themselves bore at least some of the responsibility for an antipathy between the two groups. For years, eugenics publications had taken a stance of disdain in commentary either on events involving Catholics or on publications produced by Catholics. This commentary often revolved around a range of negative stereotypes, even as it exhibited a repressed appreciation for the vigorous birthrate in Catholic communities. First, the AES had characterized Catholics as a group that held peculiar views on the significance of racial differences in society. Second, it had assumed that the intellect of the Catholic laity was dominated by the hierarchy, with no rational or scientific thought of its own. Finally, it perceived Catholic moral theology as mired in medieval methodology and antimodern rationales. Together, these stereotypes left readers with the impression that Catholics lingered in a mindless state of unenlightenment.

Accordingly, the AES executive committee turned to Ryan and Cooper, who were recognized both within and outside of the Catholic community as progressives who were interested in social policy and public welfare. Colleagues at the Catholic University of America, Ryan and Cooper were prominent members of a generation of Catholic social thinkers who struggled to negotiate the proper fit of Catholicism to contemporary America, especially in the wake of the 1907 papal condemnation of modernism.[16] The pope's condemnation doubtless had a stifling effect on Catholic intellectual life in the United States. Nonetheless, a handful of individuals, committed to the methodologies of the newly emerging social sciences, were steadfast in a difficult attempt to reconcile their faith with their intellectual pursuits. A commitment to this particular blend of developing scientific methodology and a deep concern about social issues made these priests well suited to engage with the American eugenics movement, whose faith in scientific solutions rivaled Ryan's and Cooper's own commitment to Catholic teaching.

John Montgomery Cooper's dual career as an anthropologist and as a religious educator placed him in direct contact with eugenics enthusiasts. A descendant of English Quakers, Cooper was born in Montgomery County, Maryland, and grew up in Baltimore. Educated at the North American College, by the year of his ordination in 1905, Cooper had completed a doctorate in philosophy and a doctorate in sacred theology. Shortly thereafter, the rector of the Catholic University of America invited Cooper to teach religious education to the university's undergraduates as a member of the Department of Apologetics. Trained as an anthropologist, Cooper's original fieldwork was among the Tête de Boule of Ottawa, in 1916. Over the next two decades, he traveled throughout Canada and the northern United States to observe the Ojibwa, the James Bay Cree, the Montagnais, and the Gros Ventre. Close observation of these peoples convinced Cooper of the value of inductive reasoning and of the basic unity of the human family. Additionally, Cooper took up the position of secretary for the women's committee of the National Catholic War Council in 1918. This job placed him in the middle of Catholic social reconstruction efforts and meant that he had significant contact with social reform organizations like the American Social Hygiene Association; it also meant he had to deal with the moral questions associated with prostitution, birth control, eugenics, and sex education, and that he developed an intimate understanding of the ideas of Mary Ware Dennett, Margaret Sanger, and David Starr Jordan. By 1920, Cooper had returned to his primary career as a full-time faculty member at the Catholic University of America, where he established the Department of Religious Education, served as a professor of anthropology in the Department of Sociology, and eventually founded the Department of Anthropology in 1934.[17] Considered all together, his work in religious education, anthropology, and social reform constituted an important foundation for his response to eugenics and the proposals put forth by the American Eugenics Society, making him a logical choice to participate in the advisory council that laid the groundwork for the AES in 1923.

Evidence of Cooper's thinking about marriage and family life exists in the texts that he produced in the 1920s for undergraduates in his religious education courses. His work, *Religion Outlines for Colleges*, became a standard text for Catholic students who did not intend to pursue the priesthood.[18] Composed of four volumes, the series presented students with outlines and questions that were essential to their religious training as laypersons. He urged his students to take an inductive approach to Catholic moral theology, beginning with their own experience. Hence, the volumes dealt not only with theological principles, virtues, and practices, but also with concrete daily experience—lived religion. From this perspective, he dealt with character education, relationships, sex, marriage, family, social hygiene, birth control, and eugenics. The first volume, likely composed just around the time the AES came into being, included a pronatalist emphasis, with passages that explained the social significance of marriage and parenthood to personal development and the survival of "the race." Cooper explained: "Just as death is, humanly speaking, the greatest harm that can befall the individual, so death is, humanly speaking, the greatest harm that could befall the race. And were no more children born, the race would die. Love of neighbor, therefore, in its Christian and Catholic sense, is deeply interested in birth."[19] This assertive pronatalism echoes the concerns of other social scientists about "race suicide" and reflects Cooper's worries about the influence of the birth control movement. In subsequent volumes of the series, we see the emergence of a more nuanced view of marriage and family life, less reflexive in its promotion of "the race" as humanity's ultimate priority and including not only an awareness of the range of social and cultural factors that can influence family life, but also a consideration of biological and environmental influences.

Most significantly, the fourth volume, published in 1930 and subtitled *Life Problems,* was based on concerns that Cooper's students had raised in his fifteen years of teaching religion. Drawing from the large range of questions brought to him, Cooper addressed the difficulty of finding a mate and the need to acknowledge such issues as wealth, moral character, compatibility, and health. With regard to health and the eugenics movement, Cooper instructed his readers, "While the Catholic may and does disagree with some of the proposals made in the name of eugenics by the radical left wing of the eugenics movement, he may and should be in hearty sympathy with conservative and scientific eugenics as such." In recognizing the appeal of eugenics across the political spectrum, Cooper acknowledged some of the movement's more-radical protagonists, such as Margaret Sanger and Havelock Ellis. Organized eugenics in the United States, on the other hand, was primarily conservative in its vociferous advocacy of traditional gender roles and of pronatalism for superior stock. Cooper sympathized with this element of the movement, and he counseled his students to marry into hearty, long-lived families, and to avoid choosing partners from families that exhibited feeble-mindedness, insanity, or emotional instability.[20]

While John Montgomery Cooper's intellectual interests and work made him a logical choice for the AES invitation, John Ryan's prominence in social reform causes made him likewise attractive to the organization. After an early career focused primarily on economic issues, in the 1920s Ryan sat on the boards of scores of liberal and reform organizations, from the American Civil Liberties Union to the Public Ownership League, groups that occasionally took a stand on specific issues that contradicted his own views. Ryan's biographer, Francis Broderick, has observed: "To a reforming group like the American Civil Liberties Union, he looked like a line to a conservative buoy. Non-Catholic reformers were happy to invoke, even in this oblique form, the favor of American Catholic opinion." At the same time, Ryan hoped to influence these groups' organizational agendas, and perhaps to deter troublesome initiatives.[21] No doubt this desire held true for his association with the AES as well. Even though Ryan publicly opposed much of the AES's agenda, there were elements that attracted him to the organization and pushed him to accept its invitation to join its Committee on Cooperation with Clergymen. For instance, the organization's "Ultimate Program," introduced in 1923, called for mothers' pensions, programs to encourage parenthood (among the fit), differential salaries based on family size (and quality), labor legislation to regulate the hours and working conditions for women and children, and for a general effort to improve international relations to prevent wars.[22] While Ryan likely rejected the standards by which the AES leadership created their hierarchy of fitness, all of the programs mentioned were compatible with his vision for social and economic justice.

The editors of eugenics journals and leading figures in the AES attempted to engage the more progressive Catholics, frequently turning to Ryan and Cooper for the Catholic perspective on social issues related to eugenics. Sometimes, the two men participated readily; on other occasions, they deflected these requests for a variety of reasons having to do with their schedules and their areas of expertise. In comments, Ryan was clearly more hostile to the eugenics agenda, frequently claiming that he had little to add to the discussion at hand. Cooper, on the other hand, was willing to enter into open dialogue with the eugenicists, in personal correspondence if not in the press. While Cooper never compromised his dedication to Catholic moral principles, he frequently engaged eugenicists on the methodological underpinnings of their proposals. As both an anthropologist and a religious educator, Cooper was in a unique position to communicate Catholic social teachings while honoring the experimental and statistical methods of both the social and natural sciences. Ryan and Cooper represented two of the most socially engaged, thoughtful, and well-educated Catholic public intellectuals in the interwar years. Their interactions with eugenics advocates in symposiums organized by the journal *Eugenics*—both Ryan's frustrated reticence and Cooper's eager argumentation—reveal a great deal about the conflict and convergence between Catholic social teachings and eugenics ideology on several key issues, including charity, racial prejudice, immigration, feminism, and birth control.

Beyond its committee structure, the AES also looked for ways to connect with a much broader constituency, through the nation's parishes, synagogues, and meeting houses. While the committee was interested in educating clergymen to the benefits of a eugenically conscious and eugenically structured society, their real goal was to work through priests, rabbis, and pastors to convert the hearts and minds of congregations all over the United States. By garnering the support of the nation's religiously devout citizens, the leaders of the eugenics movement reasoned, they stood a much better chance of mustering widespread local backing for their social policies. Hence, the activities of the CCC worked hand-in-hand with the initiatives of other AES committees that focused on public education.

In the March 1926 issue of *Eugenical News*, the CCC announced a contest that would offer a cash prize for the best sermon delivered on eugenics. The deadline was July 1, and the call for participants was circulated among 180 religious publications. By April, the committee had received 145 requests for the contest rules as well as for the *Brief Bibliography of Eugenics* and the *Eugenics Catechism*.[23] The AES held an annual sermon contest for the next five years and published the winning sermons in *Eugenics*. In 1930, looking to capitalize on the messages contained in the winning sermons, the AES board of directors suggested publishing an edited collection that would included the ten best sermons from that year's contest as well as excerpts from previous years' winners. They planned to consult with the Federal Council of Churches about the best way to distribute and publicize the collection.[24]

The contest constitutes the main official effort of the AES to garner the support of clergy not directly affiliated with the movement. But this method of outreach suggests that the AES had little understanding of Catholic liturgical practices and little true interest in attracting Catholic clergy to the movement. Since the mid-nineteenth century, Catholic priests often delivered only one sermon a week, at Sunday's central mass, with only brief statements at the other masses. Most priests were not known to be talented preachers, and they primarily focused their attention on doctrinal matters.[25]

In the end, the Catholic clergy were not a specific target of the CCC's outreach efforts. Had they been, CCC materials would have made a more conscious effort to bridge the significant distance in pastoral practice that existed between Catholics and Protestants. Perhaps they suspected that they were unlikely to garner favor with the nation's nearly nineteen million Catholics, based on the demographic makeup of that population. Perhaps they were only interested in attracting the attention and affirmation of the thin layer of progressive, educated, middle-class Catholics who matched eugencists' vision of racial superiority. Either way, the CCC's efforts to invite Catholics to share in their mission to reform the race was a token effort, and for the most part, an unsuccessful one.

Racial Difference and Immigration Restriction

Eugenicists, as we have seen, viewed a hierarchical ranking of human beings as inevitable and natural, framing such ranking in racial terms of purity and superiority. Catholics, not surprisingly, demonstrated a vastly different understanding of racial and social hierarchy. The gulf between these two positions created one of the most durable and long-standing conflicts between Catholics and eugenicists, particularly since, in twentieth-century America, the importance of race filtered into nearly every aspect of life. The AES pursued a policy agenda that brought these differences into sharp relief.

One example, from a near-endless rhetorical supply, of the AES's deeply held beliefs about white racial superiority is Charles Davenport's review of Earnest Sevier Cox's volume, *White America*. The founder of American eugenics extolled not only Cox's book but also the author himself, who was instrumental in securing the passage of Virginia's landmark 1924 antimiscegenation statute, "The Racial Integrity Act." Davenport gushed: "America is still worth saving for the white race and it can be done. If Mr. E. S. Cox can bring it about he will be a greater savior of his country than George Washington. We wish him, his book and his 'White America Society' godspeed."[26] Davenport's sense of the threat to racial purity and white supremacy reflected his perspective as a man who was unlikely to accept common religious commitments as more important than perceived racial differences, which was official Catholic teaching, as we shall see in chapter 6.

Davenport's thinking can be taken as a surprisingly candid but nonetheless accurate articulation of the general attitude of the AES leadership. The AES supported efforts to maintain racial purity, but the organization was also cognizant of the controversy such measures might create. For instance, in October 1925, Madison Grant raised the issue of miscegenation at the annual AES meeting. The author of *The Passing of the Great Race* (1916), Grant had been a long-time proponent of the myth that America's founders were of a biologically superior Nordic stock and that the immigrant flood of inferior races threatened to destroy that which was good and pure in the American gene pool. In the meeting, he cited the fact that mixed-race persons of African American descent were now claiming to be American Indian to subvert the terms of the Virginia Racial Integrity Act, which allowed a person with one-sixteenth Indian blood to marry a white person. Additionally, he pointed out that antimiscegenation bills were being considered in Washington, DC, Michigan, and Texas, and he urged the AES to support these efforts. In response, Charles Davenport voiced his opinion that mixed-raced persons wished to be recognized as white but that they "lack[ed] capacity and self control." However, the Eugenics Committee was wary of the repercussions of pushing for further legal distinctions among non-whites, and in the end the organization failed to throw its weight behind proposed antimiscegenation laws.[27] Nonetheless, the general tone of distain for people of color was clear. The AES promoted a very specific and well-defined racial nationalism that prized healthy, vigorous whiteness above all else.

Throughout the 1920s, the editors of *Eugenical News* periodically reported instances in which Catholics loudly professed their solidarity with non-Anglo peoples in direct contrast to the stance of most eugenicists. Such pronouncements often contained a muffled critique of Anglo-Protestant nationalism and imperialism that baffled the writers at the journal. For instance, the editors were incredulous of Catholics' motives when, in 1930, some Catholics expressed their opinion that the United States should leave the Philippines. As the *News* sarcastically reported, "At the National Eucharistic Congress of the Philippines, where 40,000 gathered in Manila last month, a resolution was adopted asking all delegates to pray for the freedom of the country from American control (which means the probable restoration of the former control by the friars)."[28] The journal simply could not comprehend that Catholics would advocate for Filipino self-determination. In the eugenicists' eyes, the only logical motivation for the statement would be to extend ecclesiastical control over the population.

But for members of the National Eucharistic Congress, their position followed from the heart of Catholicism. In addition to the lessons presented by the everyday lived experiences of the diverse American Catholic population, the church's teaching maintained that Catholics were united—regardless of racial and ethnic differences—through their commitment to a common faith and a sacramental theology. This unity through faith and ritual extended not only to new immigrants but also to the small numbers of professing African American Catholics. Ideally, acceptance of the Catholic faith and its teachings made all other differences of race and ethnicity meaningless. This, of course, was not the reality on the ground in many situations, and it by no means indicates that the church failed to recognize racial differences, especially those of custom and culture. Rather, it suggests that, at least theoretically, salvation was far more important than incidental differences of skin pigmentation. However, just as with many others, Euro-American Catholics participated to varying degrees in the oppression of African Americans, from the slave-holding practices of the Maryland Jesuits in the seventeenth century and culminating with the battles over neighborhood desegregation in the twentieth century.[29]

The eugenics movement's obsession with the clear delineation of racial difference extended beyond black and white to include people of European ancestry, meaning that one of the AES's first policy goals was the closing of the national borders. The postwar years evidenced a renewed call for isolationism from middle-class America, which resulted in an explosion of anti-Communist feeling and nativism in the early 1920s. In 1921, the United States Congress passed immigration restrictions that limited the number of persons entering the country from southern and eastern Europe.[30] Harry Laughlin, the superintendent of the Eugenics Records Office, testified before the House Committee on Immigration and Naturalization in April 1920, and was soon dubbed by the committee as the "Expert Eugenic Agent." Appointed by Representative Albert Johnson (Tacoma, WA) to study immigrants in state institutions, Laughlin had determined that there was a correlation between southern and eastern European origins and mental deficiency. His findings were published in 1922 and provided ample material to back Johnson's calls for further restriction of immigration in 1924.[31]

Similarly, even before the AES had presented its Ultimate Program roadmap to the readers of *Eugenical News,* in August 1923, AES founder Irving Fisher and the rest of the interim committee had begun to pursue a legislative agenda. In April of that year, Fisher expressed his opinion that he "considered the Immigration problem the best starting point." Madison Grant suggested that they offer their support to a bill that Albert Johnson planned to introduce in the 1924 congressional session, rather than drafting a bill of their own. This legislation, which became known as the "National Origins Act," sought to lower the quota for immigration to 2 percent of the number of immigrants, by national origin, as counted in the 1890 census, virtually eliminating immigrants from southern and eastern Europe. The bill passed with overwhelming support in the House and the Senate, solidifying and intensifying an immigration policy that already discriminated against entire groups of European immigrants based on their country of origin.[32]

Having successfully promoted the 1924 National Origins Act, eugenicists pressed on in their concerns about racial contamination throughout the decade, focusing on groups that had not been included in those restrictions—like Mexicans and Filipinos—who hailed from areas in the Americas and US protectorates. When asked in 1929 to comment for a *Eugenics* symposium on the eugenic benefits of further immigration restriction, John Ryan replied:

> I have decided that I am unable to comply with the invitation; for I regard the project of picking out immigrants on the basis of eugenic guess-working theories as not only futile but positively harmful. Aside from the obvious physical tests and tests to determine insanity or a pronounced degree of feeblemindedness, there are, in my opinion, no scientific tests of fitness that could or should be applied to the members of any race seeking admission to the United States. I do not think that we should expose prospective immigrants to the standards and prejudices of pseudo-science.[33]

It is important to note that prominent Catholics, including Ryan himself, had for years supported immigration restriction, but their support was based on labor issues, economic needs, and on the fact that social service organizations were overburdened by the large numbers of immigrants. The overtly racial calls for restriction, first by the AES and then by Congress, were what drew the protest of the church. Ryan's dismay at the notion of racially based immigration quotas echoed the opinions of many Catholics across America.[34] He did not think that the guise of science could veil the prejudices that motivated the immigration restrictions that were now the law of the land.

Despite Ryan's pointedly antiracist response to the invitation to write for the *Eugenics* symposium, the magazine's editor, C. P. Ives, asked him to comment on a similar topic later that year. Concern about immigration from Mexico and the racial differences of Filipinos caused the editor to solicit Ryan's opinion on the racial effects of intermarriage between these groups and "the superior race." Even though Ives flattered Ryan, calling him a "nationally eminent scholar," the priest refused to participate in the proposed symposium on immigration and miscegenation, citing his lack of expertise and claiming to have "no views on the subject that are worth presenting to the public."[35] His refusal virtually guaranteed that Ryan's commentary on the pages of *Eugenics* would now be curtailed. After 1929, Ryan's views did not appear in the pages of the eugenics press again.

Unlike Ryan, John Cooper consistently positioned himself as a critical voice within the movement. When asked by the editors of *Eugenics* in 1929 to respond to the accusation that eugenics ideology was based on "racial snobbery," he responded directly and critically. Cooper argued that yes, the organized eugenics movement *did* promote the "doctrine of superior races," and particularly the superiority of Nordics. He believed that this position caused "many convinced American eugenicists" to avoid affiliation with the institutional structures of the movement. While it is hard to pinpoint the "many" eugenics sympathizers who avoided the movement because of the racism inherent in AES positions, as we shall see, the willingness of other scientists to publicly call out the movement for those assumptions suggests that Cooper may have been correct in his assessment. Dedication to the ideology of racial superiority presented a roadblock to the progress of the movement, and, more importantly, it undermined the claims that eugenics was a valid applied science. Attacking the veracity of the facts supporting the notion of racial superiority, Cooper argued, "Neither the cultural nor the psychological evidence, as it stands today, is, when submitted to detailed critical analysis, sufficient or even near-sufficient to establish with any scientific probability the superiority of Nordics or of any other racial group."[36] Significantly, Cooper turned to culture and psychology to undermine eugenics claims about race. This highlights the fact that not only did he disagree with eugenic assertions about race, he thought that by making claims based solely on biology, the movement was looking in the wrong place for answers about difference.

While their public exchanges about race, immigration, and miscegenation shed light on the stance of Catholics on these issues, the staff at the AES also dealt with Ryan and Cooper's opinions within the more private, inner workings of the organization. Maintaining his stance as a eugenics insider, Cooper rarely passed up a chance to share his views with the key players in the organization. As a member of the advisory council for the AES, Cooper had the opportunity to review official statements and committee platforms that represented the guiding ideology for the American eugenics movement, frequently affording him a chance to comment on the conflict between rigorous scientific method and subjective prejudice. In his interaction with the AES in this capacity, he continued to sound a consistent voice of criticism when presented with issues addressing immigration and differential racial characteristics.

Cooper frequently explained to Leon Whitney, the executive secretary of the AES, that the racism implied by much of the organization's agenda was unscientific and repellent to those who wished to place eugenics on a sound scientific footing. On September 8, 1930, Whitney forwarded a copy of the Committee on Selective Immigration's program to each member of the advisory council. The program included an endorsement of the National Origins Act principles and called for an extension of those restrictive quotas to include the countries of the western hemisphere, with a request for adequate funds to provide for the registration of aliens and the removal of all "deportable aliens." To secure the quality of the population further, the committee proposed overseas medical examinations for immigrants to weed out undesirables before they reached America. Finally, the program called for an admissions standard that only welcomed those immigrants "who are superior to the median American in intelligence tests."[37]

Even though Cooper, like many Catholics, supported restriction in general, he was convinced that the AES's comprehensive plan was "hopelessly entangled in Nordic pre-suppositions." Invoking the memory of the Red Scare following World War I, he informed Whitney that he was doubtful, "in view of the quite possible tyrannies and injustices regarding the policy of registering alien populations." But beyond his concern for the rights of noncitizens, Cooper focused on the "emotion and chauvinism" that formed the foundation of the immigration program. He frankly explained to Whitney: "I am profoundly convinced that so long as the pro-Nordic element so largely controls the broad policies of the Society, we cannot save face in American scientific circles. We shall continually be a society based on unconsidered propaganda motives instead of upon established scientific facts. In other words so long as the Nordics are in control, we shall lack status among thinking people."[38] Using his status as an anthropologist, Cooper continually recontextualized the terms of the eugenics conversation by reminding his colleagues at the AES that the principles of rigorous scientific method required them to provide empirical data to support their policy initiatives. Without such data, their proposals were simply prejudice masquerading as science.

The stark difference between Catholic teaching and eugenic perspectives on questions related to race transcended the casual observances of the eugenics press and became a point of contention between Catholic and non-Catholic members of the AES. Time and again, the AES leadership approached Ryan and Cooper about racial issues, and Ryan and Cooper repeatedly rejected the eugenicists' racial hierarchy as unscientific. In the process, they struggled to undermine eugenic assumptions, questioning both the methodology used to derive eugenics data and the framework used to interpret it. This process placed them in a position to reject the majority of negative eugenics policies even as they continued to support the idea of social reforms that fostered strong, healthy families.

There is little evidence that either Ryan's blunt rejection of "pseudoscientific prejudice" and "eugenic guess work," or Cooper's warnings about the undermining influence of "the pro-Nordic element," caused eugenics leaders to rethink their policy agenda during the late 1920s. The voices of affirmation from within the movement were numerous and strong. More often than not, the priests' views were published and discussed in the eugenics press as a point of contrast to the dominant philosophies of the eugenics movement. Many eugenicists dismissed their objections because of the priests' dedication to a religious institution that based its authority on the notion of eternal truths. To the eugenicists, these commitments made Ryan and Cooper appear insufficiently versed in science to question the scope and methodology of the movement. Nonetheless, their willingness to engage with eugenic questions in the pages of the organization's main publications allowed Ryan and Cooper to repeatedly remind readers that Catholics who opposed eugenics policies did so with reasons that were grounded not just in religious commitment but in an assessment of the scientific foundations of the movement.

Moreover, they were not alone: a number of biologists engaged with the movement shared Ryan and Cooper's objections. As early as 1914, the Princeton biologist and eugenics supporter Edwin Grant Conklin voiced the need to consider both heredity and environment in determining the source of human behavior, in lectures that were later published as *Heredity and Environment in the Development of Men* (1915). In the 1920s, Raymond Pearl and Herbert Spencer Jennings, both Johns Hopkins scientists and eugenics supporters, critiqued the misguided conclusions being offered by the movement, writing influential pieces for the general public.[39] In a 1928 address to the Galton Society, Conklin suggested that these kinds of scientific criticisms were worthy of consideration by eugenics advocates. Though he dismissed the criticisms of the British Catholic writer G. K. Chesterton as "sentimental," Conklin took William Bateson, William Castle, Thomas Hunt Morgan, Herbert Spencer Jennings, and Raymond Pearl seriously. Each of these scientists called for some measure of restraint in the claims that could be made, based on current understandings of heredity. Conklin agreed, calling for more attention to the role of the environment, and therefore distanced himself from the more extreme fringe of eugenics devotees; but he also hedged his bets, saying that in the end, heredity held more sway than environment. Conklin concluded his remarks by explaining: "Admittedly it is difficult to decide which human traits and stocks are *best*, but it is much easier to decide which are *better* and which are *worse*. Of course we should avoid indiscriminate condemnation of whole races or classes. We should be as democratic as Mendelian Law and recognize the better wherever it occurs." Nonetheless, he argued, "we would be recreant to duty and false to truth if we should affirm that 'all men are born equal.'"[40] Thus, Conklin, and perhaps his audience, considered the criticisms of Catholics and certain scientists, but he still held fast in his confidence that heredity was the controlling factor in differentiating among peoples, even if he was willing to soften his stance on condemning entire races. Catholics offered other standards, such as culture and behavior, for considering human difference—standards that eventually would hold more sway than the biological determinism that was so compelling in the years before World War II.

Eugenic Reproduction

While *Eugenics* and *Eugenical News* diverged sharply from the typical Catholic perspective on racial difference and immigration restriction, those same publications suggested that eugenicists quietly appreciated the high birthrate among Catholics and their traditional family structure. The eugenics movement's pronatalism offered an opening to attract Catholic supporters, but the AES's stance on contraception and the sterilization of the unfit warned away those Catholics who might otherwise be attracted to positive eugenic measures. Within this contradictory nexus of race and reproduction, members of the AES nevertheless envisioned a possible, if slim, point of commonality between their philosophy and Catholic teaching.

In 1928, Archbishop Edward Hanna of Los Angeles characterized the influx of Mexicans and persons of Latin descent and culture as presaging "the peaceful entrance during the generations to come into the possession of this fair land." The editors of *Eugenical News* were in agreement with the archbishop on what they saw as the positive aspects of the Hispanic culture, but they were less enthusiastic on the idea that these immigrants were coming to possess the land. Using the Hispanic culture to attack middle-class white women, they argued that the archbishop was correct in his cultural assertions, because "home-loving, prolific peoples, among whom motherhood still remains in honor, are in the presence of a race that are permitting the home to crumble about them, who shrink from the duty of childbearing, and who are raising a generation of weaklings by showering sickly sentimentality upon carefully limited families of pampered darlings."[41] Into that one sentence, the editors had crammed a whole slate of negative stereotypes. On the one hand, the implication was that women were being dissuaded from fulfilling their racial duty by the illusive claims of feminism—that, in the words of Edward A. Ross, they were being seduced to commit "race suicide."[42] On the other hand, those women who did have children were guilty of dominating and smothering the few children whom they did condescend to raise (what Philip Wylie would later term "momism"[43])—a trend that was certain to produce a generation of pathological and emasculated men because it robbed boys of the chance to develop their own sense of rugged individualism. Grudgingly, the editors suggested that by clinging to traditional cultural patterns, these racially inferior peoples—the "sons of Italy and Portugal and Mexico"—were avoiding the pitfalls of modernity that were so disruptive of gender norms and so threatening to racial perfectionism.[44]

Implied in this commentary on birthrates was a shared perspective on questions of feminism and the "New Woman." Many Catholics and eugenicists alike were concerned about the implications of shifting gender roles for society. Not surprisingly, their responses to the challenges posed by feminism were marked by a pronounced degree of ambivalence. Feminism was associated with the birth control movement, as well as campaigns for suffrage, the equal rights amendment, and sufficient educational and career opportunities for women. As a movement, it was problematic for both Catholics and eugenicists because it threatened to upend gender relations long taken for granted and alter the centrality of the traditional nuclear family. For many American eugenicists, feminism held the threat that "superior" women would eschew their reproductive duties in favor of careers, thereby reducing the numbers of worthy offspring. As Edwin Grant Conklin told the readers of *Eugenics*, "If only by some means the better half of all women could be made to realize that the most important social service they could render to the human race would be to have seven or eight children each and at the same time the poorer half could be induced to adopt and practice this new freedom from reproduction, the problem of eugenics would be solved." Conklin's perspective reflected the investment of the AES in promoting positive eugenics policies that would encourage "superior" women to have large families in spite of the feminist notions about independence and education for women.[45] His vain wish also reveals how fanciful eugenic thinking could be, in this case utilizing the claims of feminism for certain groups while suggesting that other groups voluntarily reject such advancements.

The question of the relationship between feminism and eugenics was not an easy one for feminists either. Significantly, female eugenics supporters often echoed the pronatalist stance of male eugenicists, arguing that increased access to appropriate education, such as courses on child development and eugenics, would encourage both women and men to perform their reproductive duty by getting married and having children.[46] In taking this position, the women replicated the tendency of some first-wave feminists to advocate for social motherhood and pronatalism in order to support their claims for equal rights. This dangerous association also led them to endorse eugenics, thereby helping to sabotage the possibility of feminist bonds across races or cultures, because the movement reinforced their cultural power by prizing both social and actual motherhood among middle-class Anglo-Protestants.[47]

For Catholics, feminism threatened to destabilize the traditional patriarchal family unit and to undermine a universal pronatalism. In addressing such concerns, Cooper offered a skeptical proposal: that women who embraced education and careers instead of motherhood might be making the correct decision for society. "Should it turn out that those who are assumedly making the great refusal are doing so out of selfish unwillingness to accept the responsibilities and sacrifices of motherhood," Cooper speculated, "then perhaps the race is just as well or better off if they fail to bear and rear offspring."[48] In this less-than-pro-feminist response—he offered no alternative motives for choosing childlessness other than selfishness—Cooper succeeded in reminding the readers of *Eugenics* that they needed to be concerned with more than just biology; they should also be concerned with social dynamics and social roles. This reminder came couched in a traditional Catholic pronatalist stance, but it suggested that there might be valid, though limited, reasons for choosing a life path that did not include motherhood.

Dedicated as he was to placing eugenics on a solid experimental scientific foundation, Cooper resisted the tendency of eugenicists to hold forth on those issues surrounding reproduction that could not be addressed with verifiable empirical data. In responding to Roswell Johnson's 1929 draft, "An Ethical Code of Eugenics," which Leon Whitney circulated to the members of the AES advisory committee and which dealt with hygiene, marriage, reproduction, and divorce, Cooper tersely wrote: "If our American Eugenics Society considers its work in life as that of irresponsible propaganda I should be in favor by all means of the Society publishing Dr. Johnson's code. If on the other hand our American Eugenics Society looks upon its aims and methods as primarily falling within the scientific field, I should consider the publication of this half digested, rambling and unscientific ethical code as entirely out of place."[49] Cooper's comments indicate the sometimes surprising degree to which he felt personally invested in the eugenics movement and its future. By continually pointing to the unscientific nature of many "official" eugenics positions and platforms, Cooper demonstrated his wish to reform the ("our") AES in hopes that it might further the goals that he felt justifiably belonged to the scientific realm.

While Catholic teaching and eugenics philosophies shared in the idea of pronatalism to a certain degree, they came to an inevitable clash when eugenicists encouraged birth control for the "unfit." Catholics suspected that eugenics policies were primarily in support of a campaign for the widespread distribution of contraceptive information and technology, partly because so few of their co-religionists were among those whom eugenicists targeted with positive measures. Works by Catholic theologians negatively linking eugenics and birth control as twin false solutions to larger social problems reinforced that connection in the minds of the clergy and the laity. Charles Bruehl's text, *Birth Control and Eugenics in Light of Fundamental Ethical Principals* (1928), was the most prominent of these. However, similar views had been expressed by Thomas Gerrard in 1912, Ryan in 1916, and Henry Somerville in 1917, all of whom wrote pieces in the Catholic press linking the eugenics and birth control movements in their propensity toward ethnic and class discrimination.[50]

In fact, the association between eugenics and birth control provoked some degree of ambivalence among eugenicists, some of whom supported wide access to contraception and others of whom worried about the tendency of middle-class women to use contraception to avoid their eugenic reproductive duties. The initial stance of the AES, published by the Program Committee in 1923, called for an end to the long-standing Comstock laws that restricted the distribution of contraceptive information. Additionally, during the 1920s, prominent birth control advocates, many of whom had no affiliation with eugenics, began to frame their discussions of the social and political importance of unrestricted contraceptive information in the language of eugenic progress. For instance, in *The Pivot of Civilization* (1922), Margaret Sanger argued that "the complex problem of the feeble-minded, and the menace of the moron to human society" illustrated "the actual harvest of reliance upon traditional morality, and upon Biblical injunction to increase and multiply, a policy still taught by politician, priest and militarist."[51] Sanger's solution to that "harvest" was the adoption of a scientific program of birth control. As the decade wore on and more-conservative eugenicists reconciled their concerns about the birthrate among middle-class women and the movement's larger goals to dissuade the "unfit" from reproducing, the alliance between eugenicists and birth controllers grew stronger.[52] At the same time, Catholic opposition to the birth control movement took on a more vociferous tone as individual priests and bishops clashed with birth control advocates in New York, Connecticut, and a host of other states.[53]

Occasionally, the passion surrounding these conflicts prompted individuals who were not as obviously scientifically inclined as Cooper to enter the fray. For example, in January 1929, Dr. C. C. Little, then AES president, addressed the AES and the Eugenics Research Association at Battle Creek, Michigan. His speech, entitled "Some Obstacles to Eugenic Progress," included a section on birth control, which prompted John A. McClorey, S.J., of the St. Joseph Mercy Hospital in Detroit, to comment in an interview that the use of birth control had the potential to limit the number of geniuses born even as it reduced the number of defectives. McClorey's comments encouraged C. P. Ives, the managing editor of *Eugenics*, to put together a symposium on the subject, in which McClorey elaborated on his views and was joined by a variety of birth control advocates who also supported the eugenics movement. In his statement for the symposium, McClorey questioned the certainty of eugenicists that so-called "defectives" would produce defective offspring. In his words: "Geniuses as well as imbeciles have descended from mad parents. Saints as well as sinners have sprung from vicious forbears." McClorey also disputed the viability of Malthusian population theories, stating that the West was in danger of dramatic overpopulation. McClorey's fellow participants were quite unified in their responses to his comments. They all argued that the Catholic Church was in the contraception business too. Clerical celibacy, as well as the use of the "safe period"—the practice of abstaining from sex during the calculated period of a woman's fertility in order to create a desired space between pregnancies—provided the same chance of reducing the production of genius as using artificial forms of contraception.[54] (While the Catholic Church eventually approved the use of the "safe period" for spacing pregnancy, the method for calculating the period turned out to be incorrect, so it resulted in many unanticipated pregnancies.)

Though the exchange offered little more than stereotypes of each side of the debate, it generated a vigorous response from *Eugenics* readers. In fact, the response was so great that shortly after the publication of the first symposium, Ives set out to arrange a sequel. This time, to provide the Catholic perspective he approached John Cooper, who had kept up a long personal correspondence about birth control with Leon Whitney and the prominent eugenicist Paul Popenoe.[55] Additionally, in 1923 Cooper had written a well-received monograph entitled *Birth Control*, in which he sought to clarify church teaching on artificial contraception and the detrimental consequence of its use for the individual, the family, and the race. In that work, Cooper considered the arguments for the use of contraception, including overpopulation; underpopulation; the problems of war, poverty, and infant mortality; and the individual's economic issues and concerns for maternal health. After treating all of those issues, Cooper reasserted the church's position that marital continence—that is, the practice of refraining from intercourse during the wife's fertile period—was the only permissible way to control family size and child spacing. In making this argument, Cooper focused on social needs and context, only mentioning Catholic faith and theology in the last few pages of the nearly hundred-page work. His approach won him praise from *Eugenical News* for having "outlined the views of those who definitely take the other side."[56]

In responding to Ives's request to participate in a symposium, Cooper once again questioned whether any sufficient scientific facts existed to allow for an informed discussion of the role of birth control in eugenics. In the cover letter to his piece, he explained: "My own very strong opinion is that we really don't know anything about it, an opinion impressed on me still more deeply by the recent discussion that appeared in the March issue. Not one of the writers really had or appeared to have any detailed knowledge of the few facts we have in the field and these are not much."[57] Cooper's tone suggested that he thought that the rhetoric in the initial forum had left everyone looking foolish, particularly in the way the debate had superficially pitted religion and science against one another. Cooper's contribution to the discussion returned to his typical refrain with respect to eugenics: "Maybe, perhaps, possibly, probably, if—that is about all we can say. We do not know much about it. . . . A few more facts, scientifically established, will help much more than all our wordy views and inconclusive guesses—including this one."[58]

In contrast to this cautious response, Harvard biologist and AES advisory council member Edward M. East's contribution to the second symposium constituted a full-scale attack on the Catholic Church, one that only incidentally dealt with birth control in relation to eugenics. East focused on what he saw as duplicity in Catholic doctrine and politics. For example, he contrasted Alfred E. Smith's claims that the church would not influence his political positions in the 1928 presidential election with the lobbying effort of the church to prevent the passage of a 1929 birth control bill in the New York State legislature. In a parallel series of contrasts, he cited further evidence of the Catholic double standard: an unmarried priesthood and unrestricted reproduction among laypersons versus objections to the sterilization of the unfit and a willingness to castrate "thousands of healthy boys to furnish sopranos" for church choirs (a practice that had long since been discontinued—and had never taken place in America). In painting such a negative image of Catholicism, East drew upon images that eugenics supporters would use over and over again in answering Catholic objections to contraceptives, and demonstrated the intense antagonism that accompanied disputes about reproductive politics.[59]

Despite the wide-ranging exchange of opinions that took place in the *Eugenics* forums, eugenicists entered the 1930s convinced that they had still failed to win the support of US Catholics. Though Cooper and Ryan appeared to have some limited affinity for the goals of the movement, more often than not they objected to the direction of eugenics policy initiatives, from immigration restriction to the distribution of birth control among the poor. Cooper's writings clearly demonstrate his support for positive eugenics, but the movement's insistence on negative eugenic measures was a perpetual impediment to his full support. That is not to say, however, that eugenic policies went without debate in Catholic circles. While the notion of easy access to contraceptives alarmed the clergy, other elements of the eugenics agenda were not as adamantly rejected. As early as 1910, the conversation among Catholic moral theologians in the pages of *American Ecclesiastical Review* clearly demonstrated that a range of opinion existed with regard to sterilization for criminals, with some maintaining that the state had the power to sterilize criminals if the threat to the community was sufficient, and others arguing that such an invasive action could never be justified.[60] But by 1927 John Ryan had written a forceful condemnation of eugenic sterilization in his pamphlet *Human Sterilization*, in which he argued that calls for eugenic sterilization were motivated by prejudice and that such thinking endangered the rights of every minority group.[61] Catholic opinions were coalescing, it seems, against a rising threat; earlier speculation about the hypothetical conditions that could justify sterilization were now subsumed beneath a growing anxiety about the breadth and potential impact of negative eugenic measures.

For the most part, Leon Whitney's correspondence with Cooper indicates that from his perspective many Catholics in the United States supported the sentiments expressed by Monsignor Thomas McLaughlin of

Seton Hall when he refused to advertise the 1930 AES sermon contest, stating, "We shall do nothing towards advancing the purposes of this Society, which we consider out of harmony with the teachings of Almighty God."[62] The monsignor's response alarmed Whitney because it suggested the utter failure of the eugenics movement to educate Catholics about the benefits of eugenic social policy. Having developed a congenial relationship with Cooper, Whitney approached him in hopes of rectifying the misunderstanding. Attaching a copy of McLaughlin's remarks, Whitney explained to Cooper: "We continually get letters from Catholic priests and Catholic laymen who seem to have an entirely erroneous idea of what eugenics is. I wonder if you would not like to write an article for the Eugenics magazine showing how eugenics is in general in harmony with Catholocism [*sic*] (except perhaps for contraception), so that we may have reprints of it struck off and sent to the Catholic institutions throughout the country."[63]

Though Cooper was not opposed to the motivation behind Whitney's request, he felt compelled to clarify the reasons why so many Catholics took a stance in opposition to eugenics. He maintained that the agenda of the AES itself turned away those unfamiliar with the movement, and even some of those who were sympathetic to eugenics. Stating the issue in his usual direct prose style, Cooper told Whitney:

> Quite aside from the question of birth control, for which in the main the Society gives its approval, there are a few other highly debatable points of program which it has advocated, namely, Nordicism and sterilization. Furthermore, a good deal of its work has been characterized by a very loose number of inferences which do not seem to be justified by the data at hand. The injection however implicit of the Nordic bias into the Society's immigration propaganda and policies appears to practically all of us in the anthropological field as unwarranted scientifically. The implicit and at times explicit approval of some of the "wild" state sterilization measures appears equally unscientific, and moreover against public welfare.[64]

Reiterating his position on race and science, Cooper maintained an interest in eugenics as a tool of social welfare while rejecting the majority of the public policy initiatives for which the AES actively lobbied in the 1920s, particularly compulsory sterilization.

Despite the almost uniform reluctance of Catholics to endorse compulsory eugenic sterilization, the leadership of the AES still held out hope that they might win over Catholics to their cause. Hence, the editors of *Eugenics* put together a forum aimed specifically at creating a dialogue involving

Catholics. Running in the May 1930 issue, the forum, subtitled "A Catholic, Some Eugenicists Speak," included the prominent eugenicists Paul Popenoe, Roswell Johnson, Charles Davenport, and, finally, Samuel M. Donovan (also known as Stephen M. Donovan), the Franciscan who was one of the participants in the original exchange about sterilization carried by the *American Ecclesiastical Review* two decades earlier.[65] While the other participants in the forum used their space to argue for voluntary sterilization statutes as part of a social welfare plan, Donovan focused on the current state of Catholic thought on the issue. In the initial *AER* discussion, Donovan had speculated that compulsory sterilization might be considered lawful under Catholic moral teaching. However, by 1930, Donovan was convinced of only two things: first, that the Vatican had not, as of yet, intervened in the question; and second, that Catholic theologians were split in their judgment, though they leaned slightly toward disavowing sterilization. He claimed that the sole instance in which a sterilization policy would be justifiable would be one in which the public welfare, or "the common good of the state," could be preserved only by preventing the unfit from reproducing. Lawfulness was Donovan's primary concern. As he explained, "What is forbidden by the law of God is wrong and may not therefore be done, however desirable it may otherwise seem to be." In spite of sterilization's popularity as a policy measure, Donovan predicted that eventually Catholic theologians would reach a consensus that sterilization was a violation of moral law.[66]

Catholics and the AES after *Casti Connubii*

Donovan was correct. Just seven months after the publication of Donovan's segment of the 1930 *Eugenics* forum, Pope Pius XI (1857–1939) released the encyclical *Casti Connubii* (On Christian Marriage), effectively settling the question of eugenic sterilization for Catholics. With regard to the increased number and popularity of sterilization statutes, the encyclical states: "Public magistrates have no direct power over the bodies of their subjects; therefore, where no crime has taken place and there is no cause present for grave punishment, they can never directly harm, or tamper with the integrity of the body, either for the reasons of eugenics or for any other reason."[67] In the United States the clergy and the hierarchy came into line with the letter's teaching right away. Shortly after the issuance of the encyclical, both Ryan and Cooper resigned from their positions at the AES. Their methods for tendering those resignations reflect their distinct relationships to the organization and the eugenics movement as a whole.

Ryan took the opportunity to make a very public statement, and even released a copy of his resignation letter to the Catholic press. In that letter, he informed Leon Whitney that he assumed that his membership in the organization had lapsed since he had not paid his dues in a number of years and was no longer receiving their journals. As he explained, it was the arrival of a copy of the Society's platform, "Organized Eugenics," that prompted him to write to clarify the situation. In reviewing the platform, he explained, it was the Society's endorsement of "compulsory legal sterilization, the dissemination of birth control information and the practice of birth control for certain classes, and an increase in the number of legal grounds for divorce," that were, to him, "abhorrent for religious, moral and social reasons." On top of those grounds, clearly rooted in the teachings of *Casti Connubii*, Ryan also used his statement to attack the scientific aspects of the organization and its policies.[68]

Leon Whitney was displeased both with the tone of Ryan's resignation and with the fact that the text of the letter appeared in Catholic newspapers.

> I can fully appreciate your not wishing to be a member of our Clergyman's Committee of the Society when several of its proposals are contrary to your beliefs. For the same reason I should not care to be a member of the Roman Catholic Church, but your church is founded upon authority and eugenics upon scientific fact, insofar as it is ascertainable. And there is a wide difference. Your church quotes authorities who bolster up the beliefs you wish to hold, while eugenics wants the truth. So we shall warmly welcome any proof from you or anyone else that anything we advocate is untrue.[69]

Whitney's irritated response clearly illustrates the positivist position that scientific inquiry provides access to objective truth whereas other types of reasoning only present a subjective view of the world. Significantly, however, Whitney's own subjectivity is undeniable. In a value-laden diatribe about celibacy and birth control, he also argued in his note to Ryan that the Roman Catholic Church was "breeding in the worst way—exacting celibacy from its best and urging its people to have large families, which only the least intelligent at present attend to, with the result that your policy is making your people worse genetically every generation."[70] With that, Whitney revealed his assumptions about the ways that Catholic teaching and culture were precipitating racial decline.

On the same day that Ryan wrote his resignation letter to Whitney, Cooper drafted his own letter, in which he admitted that he had conferred with Ryan about the material contained in "Organized Eugenics." Though the tone of Cooper's letter, not surprisingly, was much more friendly than Ryan's, he too was adamant in his concerns about the unscientific tendencies of the organization:

> For the last decade I have kept up hope that our eugenics organization would in the end come around to a program in conformity with the scientific knowledge we possess and with social welfare as we understand it. Time and again during the formative period of this program I have as a member of the Advisory Committee urged, and given my rational grounds therefore, a program more in conformity with scientific and social standards. In no case that I can recall has such urging been of any practical avail.[71]

In this statement, Cooper is repeating the concerns that he had raised in response to Johnson's "Ethical Code" and to numerous past requests for comment from Ives and Whitney. While he is still willing to claim an interest in a eugenics program, he is not willing to align himself with the standard eugenics policies advocated by American eugenicists. His investment in the AES—and an image of himself as a critical voice within that organization—is clear, as is his disappointment at the intractable differences between Catholics and eugenicists.

Given their genial correspondence through the years, Leon Whitney was much sadder to see Cooper go than Ryan. In his response to Cooper's resignation, he even commented on the fact that Ryan had written what Whitney considered "a really nasty letter." Rather than disputing Cooper's points, Whitney sought to reassure him: "Your suggestions have been most helpful. I hope you will never feel that anything you have written in the way of advice as a member of the Advisory Council has been disregarded. In fact I can assure you that it has been given much more weight than a great many of the suggestions from other members, especially when you found cause to dissent from the majority opinion." Despite his assurances that Cooper's objections had always been taken seriously by the advisory council, Whitney went on to explain that he did not feel that the eugenics movement was unscientific. Such statements indicate that, while Cooper's voice had been heard over the years, his basic methodological critiques were unconvincing to committed eugenicists.[72]

The exchanges surrounding the resignations of John Ryan and John Cooper represent the end of direct Catholic participation in the AES. The interaction of the two men with eugenicists during the late 1920s was indicative of their dedication to social welfare questions and their willingness to interact with reform movements beyond the Catholic Church. Eugenics advocates were eager to claim some common ground with Catholics on a handful of issues. The most prominent of those issues was the unbending pronatalism of the American church hierarchy. Though staunch in their opposition to artificial contraception, prominent spokespersons for the church shared eugenicists' concerns about "race suicide," even as they differed on which social groups were at risk. Also, Catholic social scientists and moral theologians occasionally surprised eugenicists in their employment of scientific method and their general support of positive eugenics measures. However, even in those instances when eugenicists found something positive about Catholic positions, they tended to combine that recognition with a surprisingly persistent handful of negative stereotypes. And those negative stereotypes reinforced the image of Catholics as backward-looking antimodernists who failed to recognize the necessity for vigilance in the preservation of racial purity.

Nonetheless, the writings of Ryan and of Cooper in the eugenics press succeeded in delineating the areas of conflict and common ground that could exist between Catholics and the eugenics movement. Those areas of divergence and overlap illustrate the inextricable connection between race and reproduction across all segments of American society. Catholic teaching and eugenic principles provided their adherents with tools for categorizing human beings and governing gender relations and reproductive choices. The two-pronged nature of the eugenics movement—with positive eugenics promoting motherhood and social welfare programs that encouraged healthy families on one hand, and with negative eugenics promoting sterilization, birth control for the unfit, and immigration restrictions on the other hand—suggested that Catholics and eugenicists could achieve some cooperation. However, the negative elements of the eugenics policy agenda made the alliance between some Catholic intellectuals and the eugenicists tenuous and temporary. While Ryan and Cooper rejected the negative elements of the eugenics agenda, many eugenicists dismissed Catholic moral reasoning. In the process, both parties suggested that the positions with which they disagreed lacked scientific foundation.

In these debates about how to gauge scientific validity, as well as about the proper priorities of American society, we see the crystallization of opposing, and increasingly hardened, ideologies. With the creation of the Committee on Cooperation with Clergymen in 1923, there was an earnest—if overly optimistic—effort to engage a pan-Christian community in the issues of deepest concern to them. By the time of Ryan and Cooper's resignations, less than seven years later, eugenicists had a much better sense of the ways that even the most liberal Catholics set themselves apart from the movement—in their challenges of the eugenicists' scientific method and of their deep racial assumptions. The conflicts represented in the pages of the *Eugenics* symposiums lay out the basic terms of the local legislative battles that would take place around the country for the next two decades as increasing numbers of Catholic clergy and laypersons raised their voices in opposition to negative eugenics policies. In this way, the readers of *Eugenics* became familiar with the Catholic perspective on eugenics policies, a perspective that would soon be increasingly familiar to the public at large.

THREE

Practical Means: Catholic Strategies for Protesting Sterilization Statutes

In 1924, the same year that the federal government tightened the requirements for immigration to the United States, the legislature in the Commonwealth of Virginia passed what would become perhaps the best-known forced sterilization statute in the country. The law's passage was followed swiftly by a test case that went all the way to the Supreme Court, where in May 1927 the justices issued the landmark *Buck v. Bell* decision: laws authorizing the compulsory sterilization of the "feeble-minded" living in state institutions were legal. Never had the growing power of the eugenics movement been as clear.

Shortly thereafter, the moral theologian and priest John A. Ryan took up his pen to address the issue of sterilization from a Catholic perspective. In the resulting pamphlet, *Human Sterilization*, Ryan argued that eugenic sterilization measures were unscientific and bad social policy; but even more important, he claimed, the *Buck* decision was a clear departure from the understanding of an individual's natural rights central to Catholic moral teaching.[1] This text was the first of many pieces published by the National Catholic Welfare Conference that encouraged Catholic citizens throughout the United States to oppose eugenic sterilization laws, which in the next ten years would emerge in a flurry before state legislatures.

Concerted political action by the laity and the clergy, both as individuals and as members of organized Catholic groups, was especially important after *Buck* because the decision amplified the import of eugenics across the United States. Harry Laughlin was head of the Eugenics Records Office and a prominent advocate of involuntary sterilization for the feeble-minded. For him, the decision marked the "end of the experimental period" for eugenic sterilization laws, just as it did for Catholics determined to use the mechanisms of the US political system effectively to thwart such laws.[2] The next three years would see increased and significantly more organized efforts by members of the church hierarchy and lay activists to defeat state-level sterilization legislation, and would culminate with Pope Pius XI's 1930 encyclical condemning eugenic sterilization. Before *Buck,* Catholic theologians and social activists had struggled to negotiate the complexities and demands of the pluralist political climate in the United States. After *Buck,* Catholics built on the lessons of their early opposition and fine-tuned a surprisingly deft strategy for voicing their objection to the laws.

Catholics were clear on their own moral objections to eugenics policies, but those objections did not comprise the centerpiece of the political strategy used to lobby state legislatures. Rather, as they now fought for a policy that was supported by their moral tradition, Catholic activists worked within the social and political context of the United States' delicate religious pluralism. Thus, they created a multifaceted argument against eugenic sterilization that would appeal to non-Catholics as well—by de-emphasizing the specifically "Catholic" objections to such policies, those grounded in what might be viewed as religious reasoning. Instead, they emphasized scientific objections to the procedure, they raised legal questions about appeal and due process, and, finally, they identified social justice issues posed by the economic and racial biases against the targeted populations. Catholic activists and social thinkers then produced a set of materials to educate their co-religionists about the moral teachings involved with eugenic sterilization. Fashioned in the late 1920s, this combination of political activism and education allowed US Catholics to offer sustained and effective opposition to eugenic sterilization legislation well into the 1940s, when pressure for such measures began to wane amid the global upheaval of World War II. And, unlike other opponents in the natural and social sciences, Catholic activists, clergy, and members of the church hierarchy were able to mobilize a significant constituency to support their positions.

An Experimental Period

While there was little legislative activity on eugenics issues in the period between 1918 and 1922, the country having been lulled by the burdens of the First World War and its aftermath, the year 1923 marked a resurgence in enthusiasm for sterilization statutes. Between 1923 and 1925, twelve laws were successfully considered and adopted in states around the country. By 1927, a total of twenty-three state legislatures had enacted sterilization statutes; the new laws were concentrated primarily in states in the Upper Midwest and the Northwest, with a smaller number in New England

and the Mid-Atlantic.[3] Most of these measures focused on the amorphous condition of feeble-mindedness. From the late nineteenth century until the 1940s, the term "feeble-minded" was used to refer to persons who exhibited a range of signs of reduced mental capacity. The fundamentally vague qualifications for this condition, and its frequent association with poverty, crime, illegitimacy, alcoholism, and a whole host of other seemingly deviant behaviors, made feeble-mindedness a catchall term for individuals who failed to live up to the standards of behavior acceptable to Anglo-Protestant middle-class Americans. This act of labeling was particularly relevant for those who were determined to be "morons," a designation created by Henry Goddard that often was applied to highly functioning and educable individuals who resisted the norms of propriety. In the first four decades of the twentieth century, such individuals often were subjected to institutionalization.[4] The call for the eugenic sterilization of institutionalized persons hinged on the assumption that feeble-mindedness was hereditary, supposedly proven by eugenic family studies. The argument in favor of sterilization held that halting the progress of these hereditary lines of degeneracy would benefit both the individuals involved and the state, by allowing such persons to live without the risks or responsibilities of parenthood and by freeing the state of the fiscal implications of their care.[5]

While the American Eugenics Society had an entire committee dedicated to legislative activity, Harry Laughlin took a special interest in the progress of eugenic legislation. He took upon himself a survey of the entire field of sterilization legislation, and he vigorously advocated for further activity. In 1922, and again in 1926, he published an optimistic and exhaustive survey of the nation's sterilization statutes, entitled *Eugenical Sterilization in the United States*. Given the judicial and legislative victories that the movement was experiencing, he exhorted his readers to support a further expansion of eugenic sterilization on social and racial grounds. Laughlin explained to his readers that "inadequates and producers of inadequates in the body politic are not capable of obeying, on their own initiative, laws concerning mate selection and human reproduction. These lowest human family strains must, therefore, be taken in hand by the state, for the promotion of the general welfare."[6] Thus, in leading the charge for the passage of eugenic sterilization statutes around the country, Laughlin called for the sacrifice of the individual to the promotion of the "general welfare." This perfectionist vision of racial development foregrounded the logic whereby collective progress of the community trumps the individual's right to bodily integrity.

By the middle of the 1920s, US Catholics had yet to devise effective means of lobbying against such legislation in the statehouses. In fact, their methods for approaching legislators were haphazard and quite varied by location. In some cases, local prelates orchestrated their own methods for responding to the popular legislation. These attempts to address legislators involved the active participation of not only clergy but also lay activists—lawyers, politicians, doctors, social workers, and leaders of various guilds and clubs. In some cases, individuals took the initiative to contact the National Catholic Welfare Conference in hopes of receiving guidance from a staff that existed to work on issues that were of national interest.

The National Catholic Welfare Conference was a direct outgrowth of the church's activities during World War I. The successful effort to mobilize Catholic support for war bonds and other activities coordinated by the NCWC prompted the US bishops to seek to continue the broad aims of the organization in a new, peacetime form. In 1919, the bishops' pastoral letter authorized the creation of the "National Catholic Welfare Council," which in 1922 was renamed the "National Catholic Welfare Conference."[7] The Reverend John J. Burke (1875–1936), editor of *Catholic World* and a major figure in the formation of the War Council, was the organization's first general secretary and guided its development through the 1920s. The historian Douglas Slawson has referred to Burke as a proponent of "militant, separate-but-equal Catholicism."[8] In leading the NCWC's day-to-day work, Burke presented a strong and distinctive Catholic voice on national issues, but he recognized the need to frame the organization's perspective within present-day social and political conditions. Under Burke's leadership, the NCWC set out to provide support and guidance to American Catholics on all issues that touched the national church, extending beyond the reach of individual dioceses. Some of these issues included general parochial school policies, prohibition, public morality, and immigration.[9] Structurally, the NCWC included a number of departments that eventually had a hand in responding to the eugenics movement and its policy agenda, including the Social Action Department, which was headed by John Ryan and dealt with citizenship, industrial relations, and charitable organizations; the Press Department; the Department of Lay Activities, which oversaw the National Council of Catholic Women and the National Council of Catholic Men; the Department of Education; and the Legal Department.[10]

Thus, in 1926, when the New Jersey state legislature began considering a sterilization bill, Mary G. Hawks, president of the National Council of Catholic Women (NCCW), was able to turn to General Secretary Burke at the NCWC to seek his advice on how to craft a successful opposition strategy.[11] Hawks was eager to take a stand against the New Jersey legislation. The NCCW was the largest national organization of Catholic laywomen, and it actively pursued the church's social justice mission in both the private and political arenas.[12] Given her experience as an activist, Hawks sought Burke's approval of her plan to counteract the efforts of the eugenics policy advocates. She was part of a coalition of laymen and women that had requested a hearing at the state capitol on the bill. At that hearing, Hawks and the diocesan director of the Holy Name Society, a men's charitable and devotional group, voiced their protest against the bill. Their status as laypersons allowed them to represent the views of ordinary Catholics in a way that the clergy could not, especially given the ongoing suspicion in the 1920s of the clergy and their allegiance to Rome. In a state where nearly 30 percent of the residents were Catholic, such lay representatives were sure to be given a serious hearing.[13]

Though Hawks had engineered a way for lay Catholics to have their voices heard, she was far from certain about what exactly their message should be. Hence, she sought Burke's recommendations on constructing a convincing argument against sterilization. First, she proposed to quote from an address made by Thomas Verner Moore on the topic of the "Sub-Normal Child," in which he had argued against the scientific claim that mental deficiency was primarily hereditary. Second, Hawks meant to argue that "the citizens of New Jersey are asked to legalize a serious violation of the person and personal rights of defenceless [*sic*] citizens for an end not justified by the scientific findings."[14] Thus, she would present objections based on the grounds of scientific rigor and individual rights.

Burke wrote two letters in response to Hawks's request for help. Well aware of the secular environment in which she would give her testimony, he counseled: "While Catholic authority is the last word on the subject for us, it is only an opinion to those outside the Church. It is better to go along the line that such a proposal is inexpedient and will not accomplish the end it is designed to accomplish." Burke would provide this same advice numerous times in the next several years. He recognized that on social questions Catholics needed to offer a number of nonreligious perspectives that would resonate in the pluralistic environment of the American political arena. Thus, in his first letter to Hawks, he laid out a number of logistical and theoretical positions for why eugenic sterilization was an unrealistic policy measure. His arguments refuted the notions that sterilization was a just punishment, an effective deterrent, or a therapeutic remedy. In his second letter, Burke provided Hawks with statistics from clinical studies that had examined heredity and mental deficiency and whose conclusions suggested that mental deficiency was not primarily hereditary. Burke confirmed Hawks's instincts in urging her to base her testimony on scientific grounds rather than on the issues of individual rights or moral law.[15]

Similarly, in February 1927, when a priest in Vermont wrote to Burke asking for material he could use in his response to a sterilization bill before that state's legislature, Burke responded with a plan that emphasized the secular, political, and practical reasons for rejecting sterilization as an effective form of public policy.[16] Burke provided the priest with a brief list of clinical studies that questioned the hereditary nature of mental deficiency and a more extensive list of references to the moral issues involved with sterilization. However, rather than urging the clergyman to present points of moral argument to the legislature, Burke suggested a more practical plan that reckoned with the realities of the American political system:

> The practical means of defeating such legislation is to have medical authorities who are informed in the premises to present their arguments against to legislators. Also, it would be very advisable to have at least one capable lawyer of distinction who could address a legislative committee. The laymen, doctor and lawyer, are usually more effective than the priest, for the latter's arguments, we have found, are generally looked upon as "religious." Indirectly, Catholic organizations of men and women can do good work by voicing their own intelligent protest. If general medical organizations, or organizations of lawyers, or non-Catholic welfare organizations, could be led to voice their protest, it is always helpful.[17]

Burke's plan reflected his sensitivity to the complex political and social realities facing Catholics in the United States in the 1920s. The best way to wage the battle against sterilization legislation was to focus on scientific and legal realities in the context of the American political tradition, drawing strength from but not focusing on Catholic moral theology.

In the years just prior to *Buck v. Bell*, Catholic activists sought to influence their state legislatures as eugenic sterilization bills were introduced. Though Burke offered advice to those who contacted him about eugenics and sterilization, not everyone sought his counsel. In addition to the inquiries from New Jersey and Vermont, the press coverage of legislative proceedings contained in local diocesan newspapers indicate a flurry of activity in other states concerning sterilization statutes during the spring of 1927, just at the time the *Buck* case was under consideration by the Supreme Court. Between February and April alone, diocesan newspapers reported that eugenic marriage and sterilization legislation was under consideration in Colorado, Indiana, Maryland, Minnesota, New Jersey, Ohio, Vermont, and West Virginia.[18]

In the midst of this activity, Catholic strategies for opposing sterilization legislation reflected two distinct considerations: the political environment in the United States, and Catholic teaching on the rights of the citizen and the role of the state. In reminding his correspondents that Catholic moral teaching "is only an opinion to those outside the Church," Burke pointed to the precarious balance that Catholics had to strike between respecting the authority of the church and participating in the democratic political process. This situation had troubled Catholics in American public life for decades, as they faced periodic accusations—most recently during the spike in nativism immediately after World War I—of having divided loyalties and of being so disciplined by church structures and teachings as to be patently incapable of participating in liberal, democratic structures. Hence, as Catholics registered their objections to eugenics policy initiatives, they also struggled to remain faithful to the church's moral teachings.[19]

Though the rise of the eugenic sterilization movement was a great cause for concern, the staff at the NCWC and the Catholic bishops found themselves active on many political fronts in the period after World War I. One of the most troubling issues was the effort to federalize the nation's educational system. The National Education Association, supported by nativist groups such as the Ku Klux Klan, sought to remove control of public education from state and local governments and to create a compulsory national system. This posed a significant threat to Catholics, whose parochial schools supported and sustained immigrant-community languages, practices, and customs in a way that would not be possible if public schooling were mandated. In response to the proposed Smith-Towner Act, many bishops argued that a federal system of education would overstep the boundaries of reasonable government by failing to respect the rights of the individual and of the family as the basic unit of society. Families deserved the right to make their own decisions about the education of their children. In conjunction with this view of the relationship between the state and the citizen, Catholics argued for local control of schools and presented a positive view of parochial schools. The NCWC fashioned a lobbying plan to defeat the Smith-Towner Act that sought to coordinate the efforts of local chapters of the National Council of Catholic Women and the National Council of Catholic Men, as well as those of the 21,000 parish priests.[20] As it turned out, Smith-Towner never came to a vote before either branch of the legislature during the 66th Congress,[21] and thus the NCWC never needed to deploy its network of lay activists. Still, the organization's willingness to mobilize laypeople and clergy demonstrates a clear strategy about confronting legislation hostile to Catholic interests.

Aware of the popular perception that Catholicism was antithetical to democracy, in 1922 John Ryan wrote a book about the Catholic doctrine on church and state that reveals a good deal about the emerging strategies that Catholic activists were using to oppose sterilization statutes. In part to diffuse the suggestion that Catholicism and democracy were incompatible, Ryan carefully outlined Catholic teaching on the distinct rights and responsibilities of the church, the state, and the citizen, and their relationship to one another. Drawing on the teachings of Pope Leo XIII in his encyclicals *Immortale Dei* (1885) and *Rerum Novarum* (1891), Ryan explained that the Catholic Church and the secular state maintained separate spheres of authority, with the church having dominion over areas of moral law and the state having dominion over areas of temporal and civil law.[22]

Ryan recognized that situations could arise in which there would be areas of overlapping jurisdiction, primarily in the areas of marriage and education. Indeed, the issue of marriage—and, by extension, matters of sexuality and reproduction—and that of education proved to be significant points of controversy for Catholics in the 1920s.[23] Given these instances of conflict, an understanding of the rights and duties of both states and citizens was essential, Ryan wrote. According to Catholic doctrine, "the State is under obligation to promote the welfare of its citizens, as a whole, as members of families, and as members of social classes," but the jurisdiction of the state is not so complete as to allow it to violate a person's natural rights.[24] In his articulation of natural rights, Ryan emphasized the affinity between Catholic teaching and the political theory of rights articulated in the founding documents of American democracy. These rights—the right to life (or freedom from arbitrary physical assault), liberty (freedom of movement, education, religion, speech, and writing), and the pursuit of happiness (the rights of marriage and property)—belonged to each individual "because of his nature, because he is a person, and because of his intrinsic sacredness." Ryan maintained, "As the State does not create or confer these rights, it cannot take them away." Thus, in the United States natural rights were also civil rights.[25]

From this perspective, the state and the citizen are bound by what moral theologians refer to as "legal justice," which involves the obligation of citizens to obey the laws and to take whatever steps are necessary to participate in the securing of the common good. Ryan explained that in a representative political system, citizens are bound to perform their electoral duties in selecting fit legislators and in protesting unjust laws. Additionally, while Catholics might consult church teaching on matters of morality, their political protest or civil disobedience constitutes the activity of concerned and informed citizens who wish to promote the common good through the normal structures of liberal democracy. And such activity does not constitute interference of the church in civil matters.

The Catholic opposition to eugenic sterilization statutes echoed Ryan's articulation of the church/state dynamic. In working to construct an effective response to the campaign for eugenic sterilization statutes around the country, activists formulated political strategies that were remarkably consistent in their use of Catholic teaching on the relationship between church, state, and citizen. Moral theologians were concluding that sterilization statutes would empower the state to inflict unnecessary bodily harm on individuals, and Catholic citizens had engaged the political process to defeat those statutes. One among the vast plurality of interest groups in the United States, Catholics recognized the need to frame their arguments in such a way as to provide Catholics and non-Catholics alike with significant reasons to reject eugenics policy initiatives.

Drafting an Opposition in Ohio

Lay activism in Ohio produced the first literature drafted specifically to support Catholic opposition to state eugenic sterilization laws. During the spring of 1927 two separate bills calling for eugenic sterilization were introduced in the Ohio State Senate. The first bill sought to mandate the sterilization of degenerates prior to marriage, and the second called for the sterilization of institutionalized hereditary defectives and criminals.[26] Theoretically, these two measures would have ensured the eugenic future of much of the Ohio population. However, where eugenics supporters saw a social policy based in science, Catholics saw dubious science and a policy that contradicted their understanding of natural law and human rights.

At the end of March, Charles F. Dolle, executive secretary of the National Council of Catholic Men (NCCM), brought these bills to the attention of Ohio's bishops. He urged the heads of the Cleveland, Columbus, and Cincinnati dioceses to send "some high medical authority" and other appropriate representatives to lobby the house's Public Health Committee. In the meantime, Dolle had had a positive meeting with Ohio governor A. Victor Donahey, suggesting that the governor could be persuaded to reject the compulsory sterilization legislation.[27] In an effort to better inform the governor of the emerging Catholic position on eugenic sterilization, Dolle collaborated with John Burke at the NCWC to produce a brief on the subject. Eventually, the NCCM published the brief in the form of an "information bulletin" distributed to other lay organizations. The leadership of the NCCM, which functioned under the oversight of Cleveland bishop Joseph Schrembs (1866–1945), suggested that the pamphlet would "help our groups and others to a better understanding of this kind of legislation, a greater realization of the evil consequence which will result from it and assist them to oppose such measures."[28] Opposing sterilization legislation provided an opportunity for Catholics to act on the social justice mission embodied in their organizations and to fulfill a vision of lay activism that Schrembs had advocated since the early years of the NCWC.[29]

Organizationally, the "information bulletin" provided a blueprint for subsequent materials produced to educate the Catholic population about eugenics. First, it described both pieces of legislation in detail and then discussed the procedural problems involving the consent and appeal process associated with each bill. Second, the bulletin discussed the medical opposition to sterilization, citing the work of nationally recognized mental health physicians.[30] Third, the document provided statistical arguments contradicting the supposed inheritability of mental defects and exposing the ineffectiveness of sterilization as a means to halt the spread of venereal disease. Finally, the bulletin discussed the moral aspects of the legislation, arguing that no state of emergency existed that could warrant giving the state the power to inflict such a bodily intrusion upon its citizens. The text moved beyond strict natural law arguments to reflect on the social and cultural implications of the laws, suggesting that the vast power granted to the state through the legislation would eventually be used to oppress the poor: "Feeblemindedness is very often the cause of poverty and poverty is so often the cause of feeblemindedness that there is danger of confusing one with the other. It is upon the poor that these laws will bear most heavily. Therefore, those who have human dignity at heart ought to exercise a very careful vigilance lest under the guise of eugenic reform the rights of the poor are infringed."[31] Since the bulletin had a Catholic audience, rather than the mixed audience of politicians who would be the target of lobbying efforts, the authors took great pains to explain the objections to the bill in terms of Catholic moral teaching. The strategy illuminated the web of intricate connections in Catholic moral and social teachings, articulated most clearly in Leo XIII's *Rerum Novarum*, which linked family and reproductive issues to demands for both social and economic justice.[32]

Perhaps as a result of the efforts to educate the laity on the questions raised by eugenic sterilization, Ohio Catholics bombarded the state legislature with their views on the measures under consideration. Bishop Schrembs, claiming to speak on behalf of the "700,000 Catholic Citizens of the Diocese of Cleveland," told the members of the house Public Health Committee that "compulsory sterilization of so called feeble-mindedness in any of its degrees is a real menace to the Community and would lend itself in the hands of unscrupulous persons to the destruction of the most sacred human rights."[33] In their statement to the state legislature, Bishop James J. Hartley and Archbishop John T. McNicholas also urged caution about putting such decisions in the hands of public officials.[34] The Catholic District League of Cleveland, a local lay organization, was more forceful, condemning the legislation "as an outrage upon humanity, as a brutal interference with human liberty; as one of the worst instances of autocratic tyranny; as an attempt at barbarous enslavement of the people of this commonwealth; as a savage onslaught upon the sanctity of human personality; and as a first step towards introducing into the civilization of the twentieth century the views and usages of semi-brutes."[35] In the eyes of many Ohio Catholics, with the sterilization laws, the potential for human rights abuses at the hands of the state was simply too grave to not speak out.

On April 15, 1927, the Cleveland diocesan newspaper, the *Universe Bulletin*, reported that the sterilization bills had died in the house and that there was little chance that the bills would be revived during 1927 or 1928. The concerted effort by Catholics around the state of Ohio definitely contributed to the defeat of the bills. In addition to the statements made by the church hierarchy and the efforts of the NCCM, a variety of Catholic lay organizations had also issued statements of opposition to the legislation.[36] Hence, by the middle of April 1927, Catholics could take credit for having had a hand in defeating eugenic sterilization legislation in Ohio.

In lobbying state governments, diocesan officials and local activists were producing new materials that synthesized the opposition to sterilization from a number of perspectives. Such literature included theological and legal opinions suggesting that sterilization laws were unnecessary and impractical measures that were ripe for abuse and that exposed the disadvantaged to grave danger. Compiling these materials also forced Catholic opponents of sterilization measures to gather the scant scientific and social science literature available that refuted the claims of eugenics advocates. In this way, Catholics were able to marshal the isolated and dispersed critical objections to eugenics policies while claiming to speak on behalf of entire diocesan populations and active organizational memberships. This activism stood in contrast to that of biologists such as Raymond Pearl and Herbert Spencer Jennings, and social scientists such as Lester Ward and Franz Boas, who spoke out against the eugenics movement but were unable to mobilize a significant constituency to support their views.[37]

Buck v. Bell

Just two weeks after Catholics defeated sterilization legislation in Ohio, the eugenics movement secured its greatest legal success. On May 2, 1927, the Supreme Court issued its decision in the *Buck v. Bell* case, affirming the constitutionality of the Virginia involuntary sterilization law. Guided through the legislature in 1924, the Virginia statute provided superintendents of five state institutions with the option of petitioning the board of directors that governed each institution to sterilize inmates whom they believed would benefit from the operation. Given the successful history of constitutional challenges to compulsory sterilization statutes over the previous ten years, state officials wanted a test case for the new Virginia law. Dr. A. S. Priddy, the superintendent of the State Colony for Epileptics and Feeble-Minded, petitioned the board to sterilize Carrie Buck, in part to secure the standing of the statute with the appeals courts. The story of Carrie Buck, her mother Emma, and her daughter Vivian, provides a prime example of the zealousness of eugenics supporters and the degree to which "feeble-mindedness" served as a catchall term for social nonconformity.[38]

Four years earlier, on April 1, 1920, Emma Buck had faced a commission of experts charged with deciding whether or not to admit her to the Virginia Colony. A convicted prostitute afflicted with syphilis, Emma was the "widowed" mother of three children, some of whom were born out of wedlock. On that day in April, the members of the commission issued Emma's orders for commitment, and six days later she was admitted to the Colony. She remained institutionalized for the remainder of her life. Upon her admission to the Colony, Emma's daughter, Carrie, was placed in the foster care of J. T. and Alice Dobbs. In the fall of 1923, the Dobbs, claiming that Carrie was feeble-minded, epileptic, dishonest, and morally delinquent, petitioned to have her committed to the same institution that housed her mother.

Carrie was seventeen years old and pregnant when her commitment hearing was held on January 23, 1924. The Dobbs had failed to inform officials that, according to Carrie, she had been raped by one of their nephews. Such an omission made it easy for the institution's leadership to label Carrie as a promiscuous young woman. As superintendent of the Colony,

Priddy included "deviant" sexual behavior or "moral deficiency" on his list of characteristics warranting commitment and eventual sterilization. In his 1922–23 biennial report for the Colony, he explained, "The paroling of unsterilized, physically attractive young women from the institution (to the) best of families is not without danger; . . . it is not infrequent for them to be returned to the institution pregnant despite the best of care which was given them."[39] This assessment effectively offered a way for Priddy to erase any evidence of sexual violence that might befall women released from the Colony and to replace it with an assumption of promiscuity on the part of the women. Priddy's zealous support of sterilization would come to bear upon Carrie shortly after she entered the Colony on June 4, 1924, following the birth of her daughter Vivian. In September, Priddy testified before the State Hospital Board and Carrie's appointed guardian, R. G. Shelton, that the young woman needed to be sterilized so that she could be released from the institution to lead a safe and productive life. Estimating Carrie's mental age at nine years old, the superintendent rejected the suggestion that proper training might facilitate Carrie's release. He insisted that the only alternatives were sterilization or total confinement for her childbearing years. Given Priddy's recommendation, the State Hospital Board approved the petition for Carrie's sterilization.

With the board's approval, Priddy and other eugenics advocates in Virginia knew that Buck's case could also be a test for the constitutionality of the statute in the courts. Priddy and Aubrey Strode, the attorney for the Colony, began to pursue the case right away, gathering testimony and expert witnesses. Predictably, they solicited a deposition from Harry Laughlin. Although Laughlin never actually met with Buck, he declared that Carrie was a prime example of a low-grade moron and that her feeblemindedness was most likely hereditary—a judgment he based solely on having read her files. Additionally, Priddy and Strode secured an attorney to represent Buck throughout the appeals process.

When the matter finally came before the Circuit Court of Amherst County in November 1924, Irving Whitehead, a close boyhood friend of Aubrey Strode, represented Buck. This close personal relationship ensured that Carrie would not benefit from zealous representation on her behalf during the appeals process. At the hearing, Strode called eleven witnesses to establish, anecdotally and professionally, the mental deficiency of the Buck family. His case included Laughlin's statement and those of a host of other "experts." Irving Whitehead offered no expert testimony on Carrie Buck's behalf. In February 1925, Judge Bennett Gordon issued his decision sustaining the Virginia sterilization law.[40] Whitehead pursued the matter through the Virginia Court of Appeals, which also sustained the Virginia law, to the US Supreme Court in September 1926.

Nearly eight months later, the Supreme Court issued an eight to one affirmation of the involuntary sterilization law. Writing for the majority, Justice Oliver Wendell Holmes Jr. (1841–1935) concluded that the Virginia statute did not violate the guarantee of due process contained in the Fourteenth Amendment since it provided for public hearings and the possibility of appeal. His reasoning bears reproducing at length:

> We have seen more than once that the public welfare may call upon the best citizens for their lives. It would be strange if it could not call upon those who already sap the strength of the State for these lesser sacrifices, often not felt to be such by those concerned, in order to prevent our being swamped with incompetence. It is better for all the world, if instead of waiting to execute degenerate offspring for crime, or to let them starve for their imbecility, society can prevent those who are manifestly unfit from continuing their kind.[41]

Holmes's decision indicated that the state's need to lessen the burdens brought about by individuals unable to take care of themselves outweighed the right of individuals to procreate. Justice Pierce Butler, a Catholic, offered the lone vote of dissent, but failed to write an opinion supporting his position. His ultimate reasoning is lost to history. Carrie was sterilized on October 19 and paroled in November.

We have no indication that the NCWC Legal Department monitored the progress of the *Buck* case. On May 7, 1927, Patrick Ward, Director of the NCWC Bureau of Publicity and Information, submitted a copy of the evidence and the decision to General Secretary John Burke. This packet included the arguments from Whitehead, Holmes's decision, and excerpts of testimony from Harry Laughlin and other experts. Significantly, this collection of materials included no analysis on Ward's part, nor did it contain any analysis from William Montavon, Director of the Legal Department.[42] There are a number of possible explanations for the apparent lack of awareness about the case on the part of the NCWC. First, given the fact that the Catholic population in Virginia—2 percent of the total population in 1926—was small compared to areas in the Northeast and the Southwest, the structural apparatuses for monitoring developments in the legislature likely were lacking, and vigorous lay organizations were also likely absent.[43] Second, and more importantly, since *Buck v. Bell* was a test case, it moved through the appeals process quickly, in no small part because Buck's attorneys offered little resistance on her behalf. Unbeknownst to sterilization opponents, the primary reason for this lack of resistance from Carrie Buck's legal counsel was their collusion with Priddy at the Virginia Colony. Had Shelton and Whitehead pursued their case more vigorously, they might have sought assistance from representatives of the Roman Catholic Church, who had become increasingly successful in opposing eugenic sterilization measures. However, due to cooperation among the attorneys, Catholics were placed in the unenviable position of having to react to the Supreme Court's affirmation rather than taking an active role in the appeals process itself.

Despite being caught unaware by the developments in Virginia and at the Supreme Court, Catholics in the United States reacted quickly to the judgment. Mainstream press around the country generally welcomed the decision, following the lead of the *New York Times*.[44] The decision was not received as positively in the Catholic press. The editorial staff of Cleveland's *Catholic Universe Bulletin* noted that the decision was not surprising given the tendency of reformers to advocate against reproduction for those with mental defects. "Segregation was the alternative and should be effective in an institution even more than outside," the *Bulletin* wrote. Though theologians had been successful in thwarting laws that "could easily lead to abuse" in the past, the editors speculated that the Supreme Court decision threatened to "wake up dormant movements toward like legislation in many states and may create a new religious campaign issue."[45] This perspective anticipated the increased number of sterilization measures that would be considered by state legislatures in the wake of the *Buck* decision and recognized the volatility of reproductive issues and Catholic politics.[46] Contentious issues such as the Catholic stance on birth control and divorce only magnified this tension, and the possibility of adding sterilization to the list of political controversies must have caused some anxiety. However, the growing consensus among moral theologians and lay activists was that eugenic sterilization represented a betrayal of the individual rights that the American political tradition purported to protect.

In the May 14, 1927, issue of *America*, the publication's editors composed a strongly worded condemnation of the decision. While they maintained that the state had a duty to protect the public welfare, they argued that the movement for eugenic sterilization was a diversion from the real causes of social problems. The public welfare, they argued, was not in imminent danger from the population of "defectives" and their offspring, and options other than sterilization were available to the state: "Sterilization is only another of those alluring but fallacious shortcuts to social health which have so often led us into the bog."[47] Although the preservation of the public welfare was a worthy goal, it was not one that should be advanced by violating individual rights.

In the wake of the Supreme Court's decision, the *Catholic Universe Bulletin* commissioned and published a thorough analysis done by the Reverend Francis S. Betten, a Jesuit history professor at John Carroll University. Betten denounced the decision as sanctioning a grave transgression of the natural law that applied to every human being, not just Catholics. He argued that once the state violated the integrity of an individual citizen's body, it could eventually demand much more than the sterilization of the unfit; it could call for their termination. Betten ended his condemnation of the *Buck v. Bell* decision by painting an ominous picture: "While this is being written, there is somewhere a poor weak woman expecting her doom in agony and trembling. Strong hands are laid on her in the name of law, liberty, and civilization. She is dragged away, as once the slaves were dragged to the whipping post. She leaves the shambles a cripple, a degraded person. She may console herself by the assurance of Judge Holmes that her rights have been carefully guarded."[48]

Betten's dramatic, and explicitly gendered, image served to reinforce his message that the most helpless were destined to suffer a great injustice at the hands of the state in the wake of the *Buck* decision. His attention to the gender of the victim reflects both a sense of paternalism and a prescient foreshadowing of the dramatic ways in which women would become the predominant targets of forced sterilization in the 1930s. While women represented roughly 53 percent of those sterilized at the end of 1927, over the next five years that number would rise to 67 percent.[49] This shift echoed concerns about holding back the generational tide of feeble-mindedness, a persistent theme across eugenics texts, from Goddard's depiction of the Kallikaks to Justice Holmes's description of the three generations of Bucks.

While the Catholic press publicly condemned the court's decision, in private, representatives of the NCWC and the NCCM introduced themselves to Buck's attorney, encouraging him to file for a rehearing of the case. Charles Dolle made the initial contact with both Whitehead and Strode, offering Catholic support for the petition. According to the rules of the court, they had until June 11, 1927, to ask the justices to reconsider. The possibility for a rehearing would be slim, since one of the concurring justices would have to agree to it. Dolle persevered. Whitehead and Strode were surprisingly enthusiastic about the prospect of bringing the matter before the court again. They had no desire for a reversal of the decision; rather, they wanted Justice Holmes to discuss the principles of eugenic sterilization more fully in his opinion. In the effort to expedite the drafting of the petition, Dolle offered Whitehead $100 for his services and agreed to pay for the printing of the argument for rehearing. After consulting with the NCWC attorney, Burke supported Dolle's effort and agreed that Whitehead should take the primary public role until and unless it was necessary for a Catholic to make a statement during oral arguments.[50]

In the end, the petition for rehearing failed to include anything that stood out as a discernibly "Catholic" contribution. There were no arguments about natural rights or moral theology. Additionally, while the petition argued that the opinion rendered by Justice Holmes contradicted precedent in its interpretation of the cases he cited, the brief did not include a section refuting the testimony of Harry Laughlin or the other experts. Dolle had originally included extensive material about the expert testimony in his draft of the brief, but Whitehead rejected his suggestions. In explaining this omission to Burke, Dolle said that Whitehead "felt that since there had been no testimony produced at the hearing and he had made no reference in his argument to a divided opinion upon the subject of the propriety and utility of sterilization, he would not wish to subject himself to criticism or reproach for now bringing it before the Court for the first time."[51] There is no indication that Dolle knew that Whitehead and Strode were boyhood friends, so he had no reason to doubt Whitehead's sincerity in his explanation for leaving that material out of the final petition. Reluctantly, Dolle accepted Whitehead's unwillingness to second-guess his initial arguments in the petition for rehearing. So, while the petition did get filed, it contained none of the major objections to the decision voiced by Catholic writers, activists, and social scientists. In the end, it was not enough to convince the Supreme Court to revisit the *Buck v. Bell* decision. The request for a rehearing was denied, and Dolle's efforts slipped into obscurity. The outcome of these efforts must have been a major disappointment and frustration not only for Dolle but also for Burke, Montavon, and the rest of the staff at the NCWC. Moving forward, if Catholics wanted to prevent involuntary sterilization, they would have to intervene before laws were passed at the state level.

In the Shadow of the Supreme Court

The *Buck v. Bell* decision ended the "experimental period" in the struggle over sterilization, both for eugenics advocates and for Catholic opponents. Fortified by the affirmation of the Supreme Court, eugenics advocates made the Virginia statute a model in their efforts to implement similar legislation in the remaining states. However, just as eugenics advocates felt that the decision had marked a turning point for their movement, the Catholic opposition to forced sterilization found in it an equally valuable motivation to continue their work. Although the Catholic message about eugenics and sterilization would not achieve its full development until the definitive statement by Pope Pius XI at the end of 1930, Catholic engagement with the eugenics movement shifted its tone in the wake of *Buck*. On the national level, Catholics like John Burke had to have been frustrated by their lack of awareness about the Virginia case and a Supreme Court decision that so violated their understanding of the fundamental role of the state in safeguarding individuals rights. Many active lay Catholics, such as Charles Dolle and Mary Hawks, must have shared a deep disappointment about the ultimate outcome of the case and its potential implications for future compulsory sterilization measures. For these reasons and more, Catholics began to make a more direct, concerted, and public effort to voice their opposition to eugenics policy initiatives.[52] Though they continued to voice the same concerns about the lack of rigorous science, inadequate appeals processes, materialistic motivations, and deep injustice that they had in the past, their statements took on a more forceful quality that expressed their feeling that the public was being misled by a movement that was based on discrimination and prejudice, not on the preservation of the common good.

Despite his earlier, more discrete lobbying efforts, Bishop Schrembs of Cleveland now voiced his growing frustration with the establishment of a new eugenics organization in Ohio. Launched in 1928 by the scientist and inventor Charles F. Brush, the Brush Foundation brought a new urgency to the struggle over eugenics in that state. In a commencement address to the graduates of a nursing training school, Schrembs quipped that "formerly humanity was referred to as the human race, but according to this new foundation it may be that we are to be classified with the barnyard animals or those of the zoo." Schrembs's sarcasm, echoed two weeks later in a *Catholic Universe Bulletin* editorial, was a far more biting articulation of an old Catholic position: that supporters of eugenics policies denied the integrity and authenticity of individuals and would subject them to unjust manipulation in the name of biological improvement. Rejecting those means, Schrembs argued that the church's efforts to teach virtue and self-restraint were far more just and effective than invasive eugenics measures in motivating individuals to avoid ill-advised unions.[53]

In the wake of the *Buck* decision, leaders like the bishop of Cleveland were more willing to challenge eugenicists, but they needed to be certain that they had the means to educate the laity in support of their protest.

Therefore, Ryan dashed off the pamphlet *Human Sterilization* (1927) as an initial response to the *Buck* decision. In it he argued that the eugenics movement endangered the welfare of all minority groups: "The differences between the social inconveniences arising from the existence of too many imbeciles and that resulting from the presence of too many Negroes, Mexicans, or other non-Nordics, is a difference only of degree—possibly in favor of the imbeciles. If sterilization is no degradation of personality, no violation of natural rights, why should it not be applied to all the inferior classes that bring more inconvenience than convenience to the politically dominant *elite*?" Throughout the pamphlet, he hammered home the notion that it was prejudice and materialism that motivated the eugenicists. According to Ryan, poverty would inevitably be interpreted by eugenicists as a hereditary defect, and their methods were simply "inexpensive shortcuts" proposed by individuals who "are thinking only of comparative costs and quick results."[54]

At the same time, the leadership at the NCWC saw the need for a more substantial treatment of the issues raised by contemporary reproductive politics. At the meeting of the group's Administrative Committee in November 1928, Burke proposed the drafting of three pamphlets dealing with (1) existing state laws on sterilization, (2) the role of heredity in the "propagation of defective, feeble-minded children," and (3) the "evil results" of sterilization. The Administrative Committee agreed that the issue needed to be pursued, and at the November meeting of the American prelates the chairman of the committee assured the rest of the bishops that the NCWC would produce at least three pamphlets. Early in 1929, Burke dispatched Patrick Ward of the Press Department to secure authors for the various pieces.[55]

While the pamphlets on law and heredity were reasonably straightforward, the piece dealing with the "evil results" of sterilization stimulated a good deal of discussion among NCWC officials. Having examined an internal document that summarized current studies and opinion on sterilization, the assistant director of the Social Action Department, the Reverend Raymond A. McGowan, advised Burke that he should consider securing funds for a student to make an independent study of the results of sterilization. McGowan suggested that such a study would parallel and extend the material on negative effects contained in the most widely accepted source of statistics on the subject, Paul Popenoe's pamphlet, "Eugenic Sterilization in California" (which would later be published in a collection of essays, edited by Ezra S. Gosney and Popenoe, entitled *Sterilization for Human Betterment*).[56] McGowan's proposal highlighted the relative lack of data that was critical of sterilization. The obvious reason for this imbalance was that most of the people willing to study sterilization were committed to the eugenics agenda, as either administrators of state institutions or as members of private eugenics foundations.[57]

Given the scarcity of anti-sterilization material, Patrick Ward consulted with prominent Catholic psychiatrists and social scientists, including Thomas Verner Moore, John Montgomery Cooper, and the ubiquitous John Ryan. These men suggested that rather than producing a piece on the still ill-defined "evil results" of sterilization, the NCWC should produce two pieces, one that dealt with the Catholic position on sterilization and another that addressed medical issues. Since the Vatican had not yet given definitive instructions on the issue of sterilization, Ryan would address the questions raised by sterilization from a more secular point of view. This argument had the advantage of being one that lay Catholics could communicate easily to non-Catholics without appearing to impose their religious sensibilities on the general public.[58] Because Ward could not locate a Catholic psychiatrist with institutional experience to author the pamphlet on medical issues, he turned to Dr. Charles Bernstein, the well-respected superintendent of the Rome Custodial Asylum for the Feeble-minded in Rome, New York. Strongly opposed to sterilization, Bernstein seemed to be an excellent choice, not the least because he was Jewish. According to Ward, Moore concurred that "a pamphlet by [Bernstein] would be most effective coming from a non-Catholic."[59]

In 1930, the NCWC published the four pamphlets on sterilization in a series entitled Problems of Mental Deficiency. The series represented an expansion of the plan for opposing sterilization statutes that Dolle and Burke had developed in the NCCM "Information Bulletin," since the four pamphlets covered the key material of that initial piece, but in more depth and using authors with more direct expertise. The first pamphlet, *Inheritance of Mental Defect,* was written by Dr. Ulrich A. Hauber, a professor of biology at St. Ambrose College in Davenport, Iowa. Hauber's text reviewed the current state of biology while attempting to balance the roles of heredity and environment in explaining mental defects. The biologist took great care to describe the complicated effects of multiple factors in Mendelian genetics and the differences between hereditary, prenatal, and environmental causes for defects. The result was a picture of a very complex situation that could not be dealt with by simplistic plans for mass sterilizations.[60] Bernstein's pamphlet, *Social Care of the Mentally Deficient,* presented the logical outgrowth of Hauber's piece: that persons with mental deficiency needed to be treated with adequate care and attention in order to allow them the best chance to take advantage of their strengths. Bernstein advocated a colony system in which children and young adults could be educated and trained for productive work and potential employment in agricultural, domestic, or industrial tasks.[61] Despite Bernstein's plea for education and vocational training for persons judged to be mentally deficient, William Montavon's review of the legal state of sterilization statutes in 1930 (*Eugenic Sterilization in the Laws of the States*), the third pamphlet, proved that such laws were becoming increasingly popular with state legislators, especially after the Supreme Court's decision in *Buck*.[62] Finally, Ryan authored the fourth piece on the *Moral Aspects of Sterilization*.

The four pamphlets in the series included study guides and bibliographies for use by parish study clubs. Organized by the local chapters of the NCCW and various fraternal organizations, the study clubs helped to ensure that the adult lay population was well versed in the principles and reasoning of the church's theological, moral, and social teaching. The study clubs received enthusiastic support from their local dioceses, with one Catholic newspaper remarking: "Who should be a student of Catholic affairs or a Catholic student? Our answer would be anyone who is interested in the progress of the Faith; the victory of good over evil; the welfare of society, in other words, anyone who is not entirely self-centered and selfish." Composed of not more than twenty persons, each club provided a forum for parishioners to read, study, and discuss materials provided by the NCWC on issues such as the mass, Catholic education and history, social services, rural issues, and immigration.[63]

With the Problems of Mental Deficiency series in the works, on December 31, 1930, the Vatican issued an encyclical that granted the NCWC's recent work the weight of papal sanction. Although *Casti Connubii* (On Christian Marriage) is more widely known for its condemnation of the use of artificial contraceptives, in four short paragraphs the Holy Father provided an authoritative condemnation of eugenic sterilization that reaffirmed the opposition offered by US Catholics.[64] As we shall see in the next chapter, the encyclical was drafted in response to a number of events and conditions primarily centered in Europe. Yet, over the course of the next couple of decades, until the eugenics movement as such finally went dormant in the late 1940s, the firm articulation of a Vatican teaching helped to concentrate and energize US Catholic activism against negative eugenics. Though they would continue to argue against legislative measures, based on their understanding of science and American political traditions, clergy and laypersons could be confident as never before that their position was founded on explicit Catholic moral teaching.

Even though John Ryan's contribution to the Problems of Mental Deficiency series had been planned well before Pope Pius XI issued *Casti Connubii*, Ryan reworked his text to include the Holy Father's teaching on the subject. In reviewing the work of previous moral theologians, Ryan admitted that there had been a considerable difference of opinion over whether sterilization was intrinsically immoral or simply an unwarranted practice given the contemporary social situation. However, Ryan reminded his readers that given the pope's authoritative declaration, such differences of opinion could no longer be sustained. In addition to discussing the teaching of the encyclical, Ryan took a considerable degree of care in refuting some of the practical elements of eugenics rhetoric. He argued that popular estimates of the number of feebleminded persons and their potential danger to society had been inflated greatly, and that sterilization posed significant dangers of its own. Also, Ryan was concerned particularly with the scope of sterilization laws and their potential to include cohorts of the population whose bad behavior had nothing to do with heredity. Finally, Ryan cited several opinions arguing that sterilization was an ineffective method to deal with mental deficiency. In support of this position, Ryan quoted a passionate passage by Herbert Spencer Jennings stating that poor environmental conditions produced effects similar to those thought to have hereditary origins. "In view of the foregoing facts, particularly the Papal condemnation," Ryan summarized, "it is clear that no Catholic is morally justified in promoting either the enactment or the execution of sterilization laws. This applies to private citizens as well as to public officials and public employes [*sic*], but especially to legislators, physicians, and surgeons."[65]

The publication of the Problems of Mental Deficiency series gave Burke and the NCWC a significant weapon as they armed US Catholics to fight the eugenic sterilization movement. Instead of having to cobble together a bibliography of disparate clinical studies and moral treatises, the pamphlets consolidated and summarized that material, allowing the laity easy access to a multifaceted and forceful statement against sterilization. Contradicting the medical, legal, and social underpinnings of the eugenic sterilization movement, Catholic authors had developed a set of objections that honored their understanding of natural rights and served their unique position as a minority population within a religiously pluralist nation. Thus, the medical and scientific objections appealed to rigorous scientific method and challenged the cultural assumptions inherent in eugenics rhetoric. By enlisting a non-Catholic author for one of these pamphlets, the NCWC helped to inoculate Catholics from xenophobic accusations that their objections to the eugenics movement were the isolated response of a population outside the American mainstream. While presenting secular arguments against eugenics was important, the Catholic thinkers did not abandon their concern for the moral issues involved in the sterilization question. Pius XI's encyclical served as an affirmation of the already well-developed efforts of Catholic clergy and laity in the United States to reject eugenic sterilization as scientifically unsound and morally offensive social policy. This broad response to the eugenics movement provided a blueprint in the 1930s and 1940s for a deft strategy that Catholic thinkers would employ with increasing confidence to combat eugenics rhetoric and sterilization statutes in states across the country.

FOUR

Supreme Authorities: Catholicism and Eugenics beyond the Borders

Roman Catholic priests, theologians, and diplomats lived and worked all over the world. From these disperse locations they maintained structural and intellectual ties to the Vatican and various centers of Catholic intellectual life. For the individuals engaged in the struggle over the eugenics movement in the United States, these international ties informed and shaped their interaction with one another. During the 1930s a number of developments highlighted the degree to which eugenics supporters and Catholic activists in the United States engaged in an international conversation about eugenics. And these conversations, in turn, helped them to refine their arguments for and against eugenics within American social and political conditions. In particular, the drafting and publication of the papal encyclical *Casti Connubii* in 1930 highlighted the connections between US Catholics and the larger institutional structures of the Roman Catholic Church. As an authoritative teaching from the Vatican, the encyclical reinforced US Catholic opposition to many eugenics policy initiatives. At the same time, the far-reaching teaching raised the question of the proper relationship between the church and state with respect to marriage and reproduction in a pluralistic American society. Meanwhile, dramatic developments in eugenics policies around the world served as points of reflection for US Catholics, affording them a chance to redouble their efforts in opposing negative eugenics at home.

Like the Roman Catholic Church, the eugenics movement encompassed an international community—one made up of scientists rather than believers in a common faith. The science of eugenics had its start in Great Britain in the late nineteenth century, but quickly drew followers and spawned movements in a host of European countries as well as in a smattering of South and Central American nations. Within the eugenics community, a particularly significant relationship developed between eugenics advocates in the United States and those in Germany. Beginning as early as the 1920s, scientists from the two communities cooperated with and learned from one another in their efforts to implement eugenics legislative measures. The corpus of eugenics legislation that emerged in Germany in 1933 stands out as the most dramatic accomplishment of this relationship. The German statutes were the most far-reaching of any national eugenics plan, and eugenicists in the United States hailed them as visionary public policy. Not surprisingly, many Catholics in the United States saw developments in Germany as a disturbing preview of the course that eugenics could take in their own community if they failed to halt the passage of sterilization statutes, marriage restrictions, and other negative eugenics initiatives.

Given the international popularity of eugenics, Germany was not the only place that Catholics in the United States kept a wary eye on. In the mid-1930s, readers of Catholic periodicals became aware that eugenics advocates had succeeded in a place much closer to home—in the US territory of Puerto Rico. This development was seen as a cautionary tale for Catholics on the mainland, and US Catholic commentators used the event to emphasize their traditional understandings of the need for economic justice, rather than reproductive restrictions, to solve the question of poverty.

Given this international context, the religious and the scientific, Catholics and eugenicists, worked to adapt and interpret the events and experiences of other communities as guides for their own thoughts. Eugenicists, seeking affirmation, looked to the work of German Catholic eugenics advocates as a way to win over suspicious American Catholics. At the same time, Catholic activists, clergy, and press carefully monitored eugenics developments abroad and took that progress as a warning to be vigilant against eugenics policies at home.

The international success of the eugenics movement raised three very important issues. First, it raised fundamental questions about where the ultimate authority for eugenics policy should be placed—with the work of scientists or with the pronouncements of religious figures. Second, while Catholics welcomed the clarity that an authoritative teaching from the Vatican could offer for their interactions with eugenicists, the magisterium's claim to offer guidance for all persons, Catholic or not, revived long-standing concerns about the Catholic perspective on the proper relationship between church and state. Finally, as Catholics reflected on the alarming developments in Germany and Puerto Rico, they feared a similar escalation at home, and they used this specter of grave human rights abuses abroad to bolster their opposition to negative eugenics locally. Together, these events shaped Catholic participation in American political life during this period, as they sought to uphold the principles they held dear, including the defense of individual rights, in the face of a strong international eugenics movement.

Catholic Eugenicists

In the United States, one key element of the battle for cultural authority was the effort by eugenicists to find Catholic approval for their policies. As the most substantial critics of eugenics, and as a significant minority within the overall US population, Catholics presented a serious obstacle to eugenics policy advocates in their efforts to gather widespread support for their agenda. In the 1920s, the recruitment of John Ryan and John Montgomery Cooper as members of the AES Committee on Cooperation with Clergymen stands out as the most obvious effort of eugenicists to court Catholic support for their social reforms. Having found little success in this strategy, in the 1930s the editors of *Eugenics* searched for a supportive Catholic voice outside the United States. They found one in the German Jesuit Joseph Mayer (1886–1967).

While working as a moral theologian at the Paderborn Academy in the late 1920s, Mayer had argued publicly that sterilization was permissible under Catholic moral teaching. Then, in the *Journal of Social Hygiene*, Paul Popenoe favorably reviewed Mayer's 1927 book, *Gesetzliche Unfruchtbarmachung Geisteskranken* (The legal sterilization of the mentally diseased).[1] First and foremost, Popenoe was impressed that Mayer's book carried the imprimatur of his local superior, the Archbishop of Berlin. This stamp of approval indicated that Mayer's conclusions, as paraphrased by Popenoe—"that eugenic sterilization is, in principle, to be approved in suitable cases"—fell within the framework of traditional Catholic moral theology. Popenoe understood that with this seal of ecclesiastical approval, Mayer's text could prove to be a potentially valuable tool in winning over skeptical Catholics. By pointing to the archbishop's imprimatur, Popenoe acknowledged and made a concession to the alternate systems of authority that were important in Catholic life.

According to Popenoe's review, Mayer affirmed the basic principles of eugenics: that certain individuals were unfit for marriage and reproduction and that the church and the state had a duty to work to prevent them from engaging in these activities. Mayer discussed the morality of sterilization as a means to that end, tracing his conclusions through the thought of such giants of Catholic theology as Thomas Aquinas and Alfonso of Liguori. In the end, he argued that though eugenic sterilization programs were morally permissible, contemporary sterilization programs were unacceptable: "[It has not been] satisfactorily shown that the well-being of any state is yet menaced by the propagation of its psychopaths, the inheritance of mental diseases is still obscure, the possibilities of segregation have not been tested sufficiently, and the present technique of sterilization is not perfect and in some cases (X-rays) may even do more harm than good."[2] Thus, Mayer seemed to consider sterilization morally permissible in the abstract but thought it unwarranted under current circumstances. In his review, Popenoe emphasized Mayer's endorsement of sterilization rather than the conditions he placed upon that support.

In February 1930, Popenoe provided a translation of Mayer's article "Eugenics in Roman Catholic Literature" for the journal *Eugenics*. The piece traversed the history of Catholic theology from biblical times through the sixteenth century, arguing that the Catholic Church had an interest both in the questions raised by eugenicists and in a full investigation of "the laws of nature and of life."[3] Mayer then addressed the work of contemporary theologians with respect to eugenics and sterilization, beginning with the exchange that had been precipitated by Stephen M. Donovan's article "Circa Liceitatem Cujusdam Operationis Chirurgoae Proponuntur Dubia Nonnulla," which appeared in the March 1910 issue of the *American Ecclesiastical Review*. From that point of departure, a number of German theologians began to consider the problem, with Dr. Fritz Tillman arguing that the threat of undesirable genes to the public welfare was so great that the church and the state should have extensive power to prevent the propagation of those genes. Mayer had taken his intellectual inspiration from Tillman's work and developed his own theoretical justification for "official restraint of the blind and irresponsible propagation of hereditarily defective, anti-social criminals and also asocial psychopaths."[4]

Mayer's theory found a warm reception among a number of German moral theorists, but it also faced some criticism, most of which he dismissed. Mayer was much more concerned with those whom he could count as allies in his attempt to convince Catholics that sterilization was morally permissible. In listing the host of eminent theologians who shared his interest in eugenic questions, Mayer pointed out that much of his thought had been presented to the readers of *Central-Blatt and Social Justice*, the most prominent German American Catholic periodical. Additionally, Mayer praised John A. Ryan's *Human Sterilization* as "a mature and informative pamphlet."[5]

In a lengthy translator's note, Paul Popenoe seized upon Mayer's reference to Ryan as a chance to highlight a seemingly favorable view of sterilization from a prominent US Catholic leader. Popenoe described Ryan as "one of the official spokesmen of the Roman Catholic Church in the United States," and then explained that even though Ryan "is not convinced that the present compulsory sterilization laws are desirable, he appears to accept the principle in about the same way that Dr. Mayer does."[6] Popenoe's note reveals his motivation in providing this translation for the *Eugenics* audience; he was able to use Mayer's comments to suggest that John Ryan considered eugenic sterilization a morally permissible course of action in the pursuit of communal welfare. Similarly, Mayer's article allowed Popenoe to introduce the non-Catholic readership of *Eugenics* to a handful of seemingly respected Catholic theologians who supported eugenics sterilization, albeit in Germany.

With this, the editorial staff at *Eugenics* took a significant step toward accomplishing their goal. In fact, non-Catholics and Catholic laypersons, unaware of the general opposition of theologians and Catholic intellectuals in the United States to negative eugenic practices, could have been convinced that such tactics were morally permissible in light of Catholic social teaching. Also, since there had not yet been an official teaching on the subject in the late summer of 1930, the question was still open for discussion among Catholic theologians. Therefore, the AES and its leadership continued to hold out hope that they might win over Catholics to their plan for social advancement through controlled reproduction. By turning to the work of Joseph Mayer, with its imprimatur from the Archbishop of Berlin and his discussion of traditional Catholic theologians, Popenoe and C. P. Ives appealed to the authority of the larger Roman Catholic Church to gain the support of US Catholics for their policy agenda. Throughout the 1930s, that notion, of the global power of the church and of its moral authority to dictate social policy, would lead to a heightened tension with another notion of authority, that of scientific truth.

On Christian Marriage

When Pope Pius XI issued his encyclical *Casti Connubii* on December 31, 1930, most eugenics supporters gave up hope that they could appeal to the authority of the church in the name of their cause. The encyclical clearly stated that eugenic sterilization was a violation of natural law since it unnecessarily deprived an individual of his or her reproductive capacity and most often represented an intrusion of the authority of the state on the natural rights of the individual. *Casti Connubii* was widely distributed and was viewed as authoritative by Catholics around the world.

Sterilization was just one small component of *Casti Connubii,* a document that set out to address the conditions of marriage in the modern world. The Catholic Church had always viewed marriage, first and foremost, as a sacrament, and just as important, as an institution within the bounds of church governance. But by the early twentieth century, that institution, particularly as practiced in Euro-American communities, seemed to be in danger of crumbling. Changing gender roles and the pressures of the industrializing world were wreaking havoc on an element of life that most believed (or wanted to believe) timeless. The letter of Pope Pius XI sought no less than to arrest cultural change. *Casti Connubii* reiterated the church's position that its teaching and articulation of natural law was the proper source of governance on issues associated with marriage. Hence the text authoritatively condemned the use of artificial contraceptives, the practice of civil divorce, free love unions, abortion, and both eugenic sterilization and voluntary sterilization as a means of contraception. Since the teaching was based on principles of natural law, the Vatican intended for the positions articulated in the encyclical to apply to the entire human population, not just Roman Catholics.[7]

Though Catholics in the United States were profoundly influenced by the teachings contained in *Casti Connubii,* the state of marriage in America was by no means the key factor behind the drafting of the encyclical. John T. Noonan's *Contraception: A History of Its Treatment by the Catholic Theologians and Canonists,* as the most comprehensive text on the encyclical, explains that the Vatican was primarily concerned with trends in western Europe. Noonan argues that Catholic theologians were alarmed first and foremost by a sharp decline in the birth rate in France during the nineteenth century. This sharp decrease, matched by a lack of decline in the death rate, meant that by 1850 the French population was failing to reproduce itself. The situation did not fail to come to the attention of church leaders. In addition to the situation in France, it was Thomas Malthus's 1798 thesis on overpopulation that eventually contributed to the growth of a significant and popular movement to promote birth control in the United States and in Britain. New technological advances also contributed to the popularity of birth control, which resulted in a shift away from the traditional rejection of contraception in both medical and public opinion.[8]

The Catholic Church, along with other champions of Victorian morality, held fast against these efforts at popularization. In the United States,

Anthony Comstock waged a campaign against immorality by attempting to control the content of materials handled by the US Postal Service. In Britain, both the 1908 and the 1920 Lambeth Conferences of the Anglican Bishops condemned birth control. Likewise, in 1925, the House of Bishops of the Protestant Episcopal Church condemned the practice. The Catholic Church hierarchies in Belgium, Germany, France, and the United States also waged strong campaigns against contraception. The moral theologian Arthur Vermeersch prompted the Belgium national hierarchy to a make a forceful statement about contraception in 1909. The pastoral letter, entitled "The Duties of Married Life," inveighed against the "grave sin of onanism" and instructed priests to vigorously interrogate their penitents in the confessional about their attendance to the duties of marriage.[9] However, this united front of Christian churches against birth control was shattered in 1930 when the Lambeth Conference, citing the possibility of a "clearly felt moral obligation to limit or avoid parenthood," affirmed the use of birth control within marriage in accordance with Christian principles. While the conference condemned the use of "any methods of conception control from motives of selfishness, luxury, or mere convenience," the floodgates had been opened and the new stance was a shock to Catholic moralists.[10] Around that time, a German Catholic periodical called for a revision of the church's teaching on the subject. In reaction to these international developments, Vermeersch, who was at the pinnacle of his career as a moral theologian, guided the creation of *Casti Connubii* in Rome.[11] Though now this statement might read as tremendously conservative in its stance, in 1930 its sense of urgency was a very clear response to enormous shifts, both in individual mores and in the standards of stalwart religious and social institutions.

While the encyclical is most well known for its treatment of artificial contraception, it also directly addresses abortion, marriage restriction, and sterilization for medical, social, or eugenic reasons.[12] The letter treats these three issues in conjunction because, in the traditional teaching of the church, they all in some way fail to recognize the duty of society to accept and protect each individual, rather than to privilege the rights of one person over another. The teaching against abortion, for example, accepts the fetus as an unborn individual whose life medical professionals have a duty to preserve and protect with the same zealousness with which they attend the mother. Thus, the letter holds that the role of public authorities is "to defend the lives of the innocent, and this all the more so since those whose lives are endangered and assailed cannot defend themselves."[13]

Emphasizing the role of the state as protector, rather than as a vehicle for social control, the encyclical warns against the ways that the state might interfere with marriage. In a separate section drafted principally by the German Jesuit Francis Hurth, one of Joseph Mayer's rivals, the letter condemns any eugenics legislation that would sanction forced sterilization or prevent persons from marrying due to the possibility that they might produce defective offspring.[14] In discussing these measures, the encyclical critically singles out those "who over solicitous for the cause of eugenics, not only give salutary counsel for more certainly procuring the strength and health of the future child—which, indeed, is not contrary to right reason—but put eugenics before aims of a higher order," namely, the natural law provided by God.[15] The teaching does not condemn the goal of promoting healthy children through voluntary positive eugenics measures; instead, it only and forcefully speaks out against invasive negative policies. The letter favors the power of persuasion over the use of force in the promotion of healthy offspring, explaining: "Although often these individuals [those predisposed to have 'defective' offspring] are to be dissuaded from entering into matrimony, certainly it is wrong to brand men with the stigma of crime because they contract marriage, on the ground that, despite the fact that they are in every respect capable of matrimony, they will give birth only to defective children, even though they use all care and diligence."[16] Thus, hereditary characteristics cannot be construed as active choices for which an individual can and should be held responsible. Since heredity exists outside the realm of free will and responsible decision making, the state should not use heredity to disadvantage an individual or restrict that person's rights.

The most critical example of the state punishing an individual for hereditary characteristics occurred in situations when persons were forcibly sterilized by the state. In no uncertain terms, *Casti Connubii* denied the right of the state to act in this manner, stating forcefully: "Public magistrates have no direct power over the bodies of their subjects; therefore, where no crime has taken place and there is no cause present for grave punishment, they can never directly harm, or tamper with the integrity of the body, either for the reasons of eugenics or for any other reason."[17] This teaching, that the state has no right to inflict physical punishment on the innocent, places involuntary sterilization for the so-called unfit out of the bounds of moral conduct. It does, however, suggest that the state might inflict some type of bodily intervention as a punishment for criminals.

The teachings articulated in *Casti Connubii* represent one element of the struggle for authority that marked the church's relationship with modern nation-states in the nineteenth and twentieth centuries. Though individuals might be citizens of particular political entities, their rights and responsibilities derived from their personhood, not their political status. Marriage was the manifestation of a basic human need; therefore it was subject to natural law, not civil law. Viewing itself as the logical interpreter of natural law, the church spoke on marriage in an effort to reclaim the sacramental union from the corruption of modern philosophies and ideologies, from both the growing permissiveness in individual choice and the growing efforts of interest groups to control those choices.

The Reception of *Casti Connubii* in the United States

Upon its release, *Casti Connubii* was promulgated throughout the United States by traditional methods to the clergy and the laity. The NCWC News Service distributed the text to diocesan newspapers and Catholic periodicals in January 1931.[18] In addition to appearing in diocesan newspapers and other publications, under an arrangement organized by the Catholic Truth Society of Oregon, the encyclical was broadcast on a powerful Northwest radio station. Read in fifteen-minute intervals over six consecutive Tuesday evenings, the encyclical was presented without commentary, except for an introduction, "so that the radio audience will understand that they are listening to Pope Pius XI, not his voice, but to his actual words."[19] In a similar attempt to familiarize the Catholic public with the teachings of the encyclical, the NCWC News Service reported that a copy of the text of the letter was being made available to each of the approximately one thousand families belonging to the Shrine of the Sacred Heart parish in Northwest Washington, DC.[20] Through these various efforts at distribution, at least some lay Catholics in the United States became familiar with the text of the encyclical.

The impact of *Casti Connubii* was nearly immediate. Published in many local Catholic newspapers in January and February 1931, its critique of eugenics began to appear in the writings of Catholic thinkers soon after. Writing that spring, John Ryan in particular served as a crucial bridge between the language of the encyclical and the US context for members of the clergy. When the *AER* published the text of the encyclical in March 1931, it was followed by a commentary from Ryan in which he singled out the elements of the encyclical that he felt were the most relevant for the American Catholic community. Ryan explained that in light of the teachings of the encyclical, sterilization was "authoritatively declared to be intrinsically wrong."[21]

As we shall see in chapter 5, the encyclical's authoritative teaching pushed Catholics to be more vigorous in their opposition to state sterilization laws. The Catholic clergy welcomed the pope's teaching as a clarification of the complex issues associated with marriage in industrializing nations. Though the teaching was profoundly conservative, it provided a foundation from which Catholic activists could articulate a distinct vision of their position on marriage and the family, especially since they were faced with the disruptive forces of a growing economic depression and an ever-expanding popular culture that often challenged traditional beliefs. Additionally, the encyclical bolstered the nearly unanimous opposition of Catholic moral theologians to negative eugenics policy initiatives. The reaction of the Catholic laity is more difficult to capture, but if there were dissenters to the teaching, they were not vocal in the Catholic or secular press.

In the non-Catholic press, reception of the encyclical was mixed. In anticipation of the encyclical's release, prominent newspapers, including the *Atlanta Constitution*, the *Chicago Daily Tribune*, and the *Los Angeles Times*, carried an Associated Press story that focused primarily on the forthcoming teaching on divorce, birth control, and interreligious marriage.[22] The headline from the *Los Angeles Times*, "Pope Strict with Cupid," suggests a not uncommon condescension toward the encyclical; the article noted that the Vatican would offer "no concession to modern trends." Within a few days, the Associated Press, whose coverage ran in many newspapers reaching Catholics and non-Catholics alike, offered another article whose extensive quotations from the actual encyclical provided a summary of the basic teachings and the condemnation of forced sterilization.[23] There were also hints of how the encyclical was being received by America's Protestant denominations. The *Atlanta Constitution* provided its readers with a peak into the pews of the Peachtree Christian Church by covering the Reverend Robert Burns's sermon on the teachings of the encyclical. Burns praised its emphasis on the sacredness of marriage, but based on his reading of scripture he took issue with its teaching on divorce.[24]

Beyond this commentary, the *Chicago Daily Tribune*'s Edmond Taylor provided an interpretation of the letter that again emphasized the distinction between the trends of the modern world and the teachings of the encyclical. The article's opening sentence set the tone: "In a 16,000 word encyclical that leaves believers no compromise between damnation and complete acceptance of the inflexible ideal of Christian marriage as laid down by St. Augustine in the fourth century Pope Pius XI today turned his back on the wayward modern world and took his place among the great fighting popes of the church." Taylor went on to outline the key elements of the encyclical, commenting, "Many students of church history declare it is the most important document in the field of pure morals since Gregory VII enforced celibacy on the clergy in the eleventh century." While Taylor treated the teaching on sterilization, birth control, divorce, and mixed marriage as an expected outcome of the letter, he zeroed in on the suggestion that civil law should support the church's stance on these issues, arguing that the encyclical "suggests a degree of cooperation between the church and civil authorities almost unknown since the middle ages, and deplores the too rigid separation of church and state which allows many acts, considered grave sins by religion, to go unpunished by temporal authority."[25] This aspect of the encyclical would catch the eye of other non-Catholic readers as well.

Among those who noticed were the editors at *The Christian Century*, who were convinced that rather than establishing new or surprising teachings, the encyclical revealed disturbing things about the Roman Catholic Church. In the view of the editors, "The most significant feature of the encyclical is its exhibition of the characteristic attitudes of the Roman Catholic Church with reference, first, to the subordination of individual judgment to papal authority, second, to its method of dealing with social problems and deciding upon social policies, and third, to its demand upon the state for the support of the church's decision and the pope's commands by civil laws and penalties."[26] Rome appeared to be trying to interfere with the state's power to enact social policy. This question turned on the authority of the state and the individual to reach conclusions regarding marriage, birth control, sterilization, and other issues. From the perspective of the editors at *The Christian Century*, the pope—distant, celibate, aged—could not be less qualified to speak on marital morality, particularly with his insistence on using theological rather than sociological methods to address such questions. The editors rejected the pope's call for civil governments to cooperate with the church to eliminate the "evils" that plagued modern marriage. This objection was especially resonant in the United States, amid the repeated struggles over Catholic participation in the nation's political life. With respect to these conflicts, the editors closed their commentary by speculating that "it must be highly interesting to those Catholics who have been assuring us that the church is forever out of politics and no longer contemplates connections between church and state in non-Catholic countries, to see the pope earnestly exhorting *all* who hold the reins of power to promote the union and association of the two 'supreme authorities.'"[27] Here, and elsewhere, the Vatican's desire to check the spread of practices it deemed immoral left a distinctly negative impression of the encyclical among some non-Catholics and the sense that it was not just out of touch with the times but that it was even dangerous for twentieth-century America political life.

In the midst of this kind of reception from committed Protestants, one could assume that eugenicists would uniformly condemn *Casti Connubii.* However, because eugenics supporters and most Catholics agreed on the importance of strong, prolific, nuclear families (at least within the "beneficial" segments of society), the editors at *Eugenical News* approached the letter with a conciliatory tone. They praised the encyclical, suggesting that the pope's "pronouncements and admonitions on the sanctity of marriage" would find a "hearty response in all circles which value virtue and human happiness." Despite this respect for the pope's intentions, the editors went on to criticize the encyclical's total disregard for the "God-made laws of nature." The editors accused the pope of asserting ecclesiastical authority over experimentally derived scientific truths. Here we see a curious similarity, as the eugenicist concept of "laws of nature" sounded similar but in practice was completely different from the pope's vision of "natural law." However, rather than striking a divisive note, the editors suggested that eventually the Catholic Church would come to accept the science of eugenics as a guide: "Perhaps in the not too distant future some Pope will commend the demonstrated laws of human heredity—the God-made laws of nature which govern human heredity—to the close study and obedience by the faithful; that the Church may give its blessing on all efforts by one generation to insure sounder physical, more capable mental and higher emotional and spiritual qualities in its posterity, and will point out how these sacred laws of nature which govern human heredity can be used most effectively to such an end."[28] Instead of condemning the magisterium for trying to interfere in civil matters or for being ignorant of human experience, the *News* chose to focus on the ways that the moral authority and persuasive power of the church might benefit the cause of eugenics, should Catholics accept eugenic science and principles.

The writers at *Eugenics* took a different view, clearly perceiving *Casti Connubii* as challenging many of the principles of the eugenics movement. In that publication's legislation column, Roswell Johnson speculated that the encyclical would likely increase the campaigns against sterilization statutes. Kenneth C. MacArthur, author of the journal's religion column, was more specific about the ways that the encyclical distanced the Catholic Church from the eugenics agenda. He explained that the letter placed the church "in definite opposition to various measures which by common consent of eugenicists tend toward racial improvement." Furthermore, MacArthur suggested, "these pronouncements would seem to be embarrassing for those Catholic teachers who have given at least qualified endorsement to some of the principles condemned by the Pope."[29] He viewed the church's teaching as an external silencing of Catholics who had been in conversation with the eugenics movement. He failed to acknowledge, however, that even the Catholics who had affiliated with the eugenics movement, such as John Ryan and John Montgomery Cooper, for the most part were consistent and vocal opponents of negative eugenics policies, particularly forced sterilization.

Thus we see that the reaction of Americans to *Casti Connubii* was neither uniformly positive nor uniformly negative, even among eugenicists. While eugenicists resisted the pope's condemnation of their most aggressive negative policies, they could recognize the common ground they shared with Catholics on the importance of strong and fertile marriages, even though Catholics had a frustratingly broad vision of who should have big families. Others focused on the teaching on divorce, seeing in it a distinctly antimodern stance. However, for some non-Catholics the pope's encouragement of the state to support his teaching through compatible laws appeared as an attempt by an unauthorized foreign power to intrude in the lives of US citizens. Their objections to the pope's attempt to speak for all persons about the institution of marriage highlighted the international character of institutional Catholicism and heightened their suspicions about both the Vatican and politically active Catholics in the United States. Questions about the pope's claim to universal teaching and his influence over Catholic political activism on the issues of marriage and eugenics were only the latest examples in a long-standing and much larger conversation about the separation of church and state in American public life, and the possibility that Catholicism might not be fully compatible with those principles.

Transatlantic Collaboration

The international character of the Roman Catholic Church became more visible with the release of *Casti Connubii*, but supporters of the eugenics movement also maintained a strong web of international contacts. From its inception with the work of Sir Francis Galton in England in the late nineteenth century, professional and amateur scientists interested in the racial and nationalist goals of improving populations through purposeful breeding saw themselves as part of a global scientific community. In the aftermath of World War I, this community came together in New York City at the American Museum of Natural History in September 1921 to discuss the impact of the war on racial progress and to plan for a decade of research and legislation that would lay the groundwork for eugenic activity around the world. Among the American eugenics luminaries in attendance were Alexander Graham Bell, Henry Fairfield Osborn, Madison Grant, Harry Laughlin, and Lothrop Stoddard. In addition, there were also representatives from France, England, Italy, Belgium, Czechoslovakia, Norway, Sweden, Denmark, Japan, Mexico, Cuba, Venezuela, India, Australia, New Zealand, San Salvador, Siam, and Uruguay. (Scientists from Germany and Russia were absent due to international tension focused on those two countries during the postwar period.) The wide number of countries represented suggests the worldwide popularity of eugenics.[30]

Even though the Germans had been excluded from the 1921 meeting, an especially close and admiring relationship eventually formed between them and the Americans. The American eugenics leaders Charles Davenport, Harry Laughlin, and Paul Popenoe eagerly communicated with the key members of the German eugenics movement throughout the 1920s and 1930s. Fritz Lenz proved to be particularly important in this rapprochement. As the author of a major German eugenics textbook, Lenz developed warm relations with the staff of the Eugenics Records Office at Cold Spring Harbor. Lenz also had a close relationship with Popenoe, who translated his work for US audiences and frequently reported on developments in the German eugenics movement. Lenz and his colleagues in the United States believed that the Germans may have lagged behind the Americans in legislation but that the two groups were united in their enthusiasm for the eugenics cause. Beyond the personal relationships that existed between US and German eugenicists, the Germans also received significant institutional support for their work from US sources. In the 1920s, the Rockefeller Foundation of New York supported the work of a number of German eugenicists and played an indispensable role in the establishment and funding of both the Kaiser Wilhelm Institute for Psychiatry and the Kaiser Wilhelm Institute for Anthropology, Eugenics, and Human Heredity.[31] Based on these personal and institutional ties, German and American eugenicists pursued their goals of shaping their national populations through negative eugenics legislation.

The American precedent heavily influenced the crafting of the German "Law on Preventing Hereditarily Ill Progeny," which passed on July 14, 1933. In the campaign to institute a sterilization law, beginning in 1923, scientist after scientist had turned to the US example for evidence of a strong sterilization movement. That year, in the face of political opposition, the head of the Reich Health Office made inquiries in the United States concerning the legal and scientific basis for sterilization. Although

US statistics revealed there had been a lull in the number of sterilization procedures, the German literature on sterilization nonetheless continued to look to the United States as a model for their own program. Some proponents in Germany even contradicted the findings of the Reich Health Office about the prevalence of sterilizations in the United States, suggesting that sterilizations were in fact increasing there in the late 1920s. Paul Popenoe and Ezra Gosney's 1929 study of sterilization in California provided the Germans with a wealth of statistical data about the policies in that state. As interest in sterilization rose in the latter part of the decade, the Germans turned directly to the work of Harry Laughlin at the Eugenics Records Office. The magazine *Archiv für Rassen-und Gesellschaftsbiologie* published an article by Laughlin that trumpeted eugenic sterilization as a well-accepted element of American public policy that, in combination with education, marriage restriction, and other eugenics policies, would result in the improvement of the national population.[32]

Fortified by the American example, the Germans passed a series of eugenics laws during the first year of Nazi leadership in 1933. This corpus of legislation demonstrates that the Germans envisioned their movement as encompassing the same complex vision for social development and progress that their counterparts in the United States supported. First, the German legislature passed the Law to Reduce Unemployment, to return men to the workforce and to discourage female employment through state-sponsored work and occupational training. Next, through the Decree for the Granting of Marriage Loans, the Germans provided marriage subsidies for non-Jewish couples who had been deemed free of mental or physical diseases. Having arranged to subsidize Reich-enhancing marriages, the Nazis then supported rural settlements for families with the Hereditary Homestead Law and the Law for the New Formation of the German Farmerstock. Then, in the Law on Preventing Hereditarily Ill Progeny, the Germans complemented the new marriage-loan laws by providing for the sterilization of persons with a variety of mental and physical conditions. Finally, the Law against Dangerous Habitual Criminals allowed for sterilization and castration as punishment for criminals.[33]

This disturbing complex of eugenic laws, which were the precursors for widespread euthanasia during the Holocaust, developed directly out of the international community of eugenics advocates. The close relationship between American and German eugenicists allowed the two most egregious incarnations of negative eugenics policies to feed off one another—both structurally, in that they shared legislative strategies, and socially, in that they shared one another's enthusiasm. In the United States, eugenicists supported the Germans in their legislative efforts throughout the 1930s. Even as the Americans became increasingly alarmed by developments in Germany, eugenicists continued to praise the National Socialists for their efficiency and foresight.

The significance of the Nazi eugenics legislation did not go unnoticed by Catholics in Germany and around the world. Some Catholics in Germany were implicated in the support for German nationalism and religious anti-Semitism, and the historical record suggests that the institutional church offered little in the way of opposition to Nazi policies during the Third Reich.[34] However, despite this glaring moral failure, German Catholics did challenge the biological assumptions of Nazi racism when those assumptions threatened the church's jurisdiction over marriage and the family. According to the historian Donald Dietrich, German bishops lent their support to the goals of racial purity but argued that the means for achieving such purity should not involve immorality or violence. Therefore, many German Catholics rejected much of the medical and biological reasoning associated with Nazi racial policies. Theologians such as Otto Schilling and Karl Frank rejected birth control and sterilization as means for achieving eugenics goals because they violated personal and sacred rights and because they attempted to supersede moral reasoning as a basis for the eugenic improvement. In addition to this focus on the role of moral reasoning, German Catholics argued that race was in fact a cultural rather than a strictly biological concept, and that despite their importance, notions of race and nation could not trump divine law.[35] These positions heightened the tension between Catholics and the National Socialist state. As the 1930s continued, Catholic theologians emphasized the importance of the common good and a sense of community that would protect the rights and moral development of the individual. Such a focus tended to contradict the ruthless racialism of Nazi policies, if not directly, then at least implicitly.[36]

Despite the tendency of some of the German Catholic hierarchy to resist Nazi racialism, and despite the direct condemnation of eugenic sterilization contained in *Casti Connubii*, Joseph Mayer continued in his support for eugenics policies. In addition to supporting eugenic sterilization, Mayer defended abortion and eventually euthanasia as public policy measures that contributed to the common good. Interestingly, despite his long-standing rivalry with Francis Hurth, who drafted the section on eugenic sterilization for the 1930 encyclical, Mayer was never corrected or reprimanded publicly by the church for his views. Not only that, but he also freely drafted a memo for Hitler that provided a theological opinion on the permissibility of euthanasia. Since Mayer's memo contradicted the teaching of the magisterium, his support allowed the Nazis to counteract their Catholic critics.[37]

In the United States, Catholics became aware of the German situation and the various responses to it through widespread coverage in the Catholic press.[38] Anticipating the implementation of the German sterilization law on January 1, 1934, Pope Pius XI made sterilization the focus of his Christmas Eve address in 1933. Though he did not specifically refer to the Nazi legislation, he restated the message of *Casti Connubii*, declaring that the state had no direct power over the bodies of its citizens and instructing Catholics that the church had already spoken definitively on the issue of sterilization.[39] The editors at *America* noted the pope's position, praised him as being "ranged on the side of the weak, otherwise defenseless citizens of the world," and declared the pontiff to be "the last strong barrier against the encroachments of the omnipotent State."[40] Despite their confidence in the pope's influence, the editors at *America* would learn during the 1930s that the church's condemnation of sterilization had had little practical impact in Germany, where approximately 375,000 people were forcibly sterilized between 1933 and 1939.[41]

Given the obvious similarities between the legislative initiatives in Germany and the United States, Catholic writers were quick to point out the hypocrisy of Americans who opposed the government of the German National Socialists but who supported eugenic sterilization in the United States. As Patrick Ward, director of the NCWC Bureau of Publicity and Information, pointed out, "Inconsistently, many who are quick to challenge and criticize the constitution and polity of the new German Reich as heedless of individual rights and as an unwarranted interference with the freedom of the individual, making him entirely subservient to the will and purpose of the State, are quite willing to accept and approve the action of the German government, in the matter of sterilization, as right and justified." Ward found this willingness to implement sterilization legislation in the United States to be a direct contradiction of the nation's stated commitment to individual, inalienable human rights. In his reflections, Ward's concern extended beyond the situation in the United States to acknowledge the global appeal of eugenics. Horrified by the suggestion of a McGill University professor that "totally imbecile" children should be euthanized at birth, Ward inveighed, "Today the people of every land are at the mercy of groupies and of faddists and experimenters of all kinds,—in economics, sociology, politics, medicine,—nothing is sacred to them, not even human rights."[42]

In February 1934, the Catholic physician James J. Walsh (1865–1942) also reflected on the medical issues associated with Germany's sterilization laws. Drawing on a wide variety of medical literature, Walsh argued that the German legislation relied on a faulty understanding of heredity and disease. In making this argument, Walsh trod the same path as those Catholics before him who had offered objections to eugenics policies in the United States—men like John Ryan and John Montgomery Cooper. For example, Walsh pointed to the findings of the superintendents of institutions for persons with epilepsy, that the disease resulted primarily from events during pregnancy and not before conception. Similarly, he argued, throughout the nineteenth century many physicians thought that tuberculosis ran in families, until the discovery of the tubercle bacillus had proved them wrong. Given this track record regarding what had been perceived to be hereditary diseases, Walsh concluded, "Physicians still continue to be human and therefore will undoubtedly err, and how the Germans can hope to avoid mistakes of this kind under the administration of this new law is indeed hard to understand." In Walsh's opinion, the probability of such mistakes occurring was increased by the wide range of conditions that were associated with the German understanding of the term "feeblemindedness." He feared that the German statute could "readily be abused in its application to members of political parties who happen to disagree with those in control of the government and whose disagreement is the strongest evidence of their lack of intellectual capacity." Walsh found this potential for abuse to be unmitigated by the statutory safeguards of the German Eugenics Court. The marked uncertainty of the hereditary transmission of disease suggested that the legislation was ripe for political manipulation and human rights violations.[43]

Walsh's concerns proved to be valid as events unfolded in the subsequent months of the National Socialist regime. In July 1935, the editors at *America* noted that the Germans proposed expanding their sterilization statutes to include persons with several categories of psychiatric disorders. They reported that this change in the German law had been strongly criticized at the convention of the Swiss Psychiatric Society by Dr. M. Minkowski, a highly respected Swiss psychiatrist. Minkowski argued that some of the psychiatric disorders singled out by the German sterilization law were impossible to diagnose in a biological sense, and that there was scant evidence of any generational link for those who suffered from schizophrenia. In pointing out this international criticism waged against the science of eugenics and sterilization, the editors of *America* reminded their readers of the debate surrounding such issues within the global scientific community.[44]

As the German interest in eugenics continued to demand the attention of the press and physicians around the world, Pope Pius XI again raised the issue in 1935. His sentiments were reported in the pages of the National Catholics Welfare Conference's journal, *Catholic Action*. In addressing the convention of the International Hospital Association in Rome, the pope spoke both to the demands of a group of German doctors to place the questions of eugenics and sterilization on the association's convention program and to the larger German sterilization initiative. Arguing that medical science was not the last word on the issue of sterilization, the pope counseled his audience to consider the role of human rights in the discussion. Then, addressing the request of the German doctors to include sterilization as a topic at the conference, the pope issued a clear warning to the Third Reich: "We must express the conviction that if such practices were accepted by people, states, governments, if they should enter into the practices of life, if, in a word, they should be adopted, then Our duty suggests to Us that as Supreme Pastor, We shall have to use every means of protest." As France was scheduled to host the next International Hospital Association conference, the French representative at the Rome conference expressed his support for the pope's position on sterilization and then informed the participants that while France was happy to host the next conference, it had no interest in having eugenics or sterilization on the program. Accordingly, the participants agreed to exclude those subjects from the next meeting.[45]

Even though the Catholics at the International Hospital Association agreed not to address the issues of eugenics and sterilization, the implications of the German laws were still very much on the minds of Catholics in the United States. The concern was so great in the wake of the pope's repeated condemnation of sterilization that in July 1936 the *AER* carried an extended discussion of the morality of the German sterilization statutes. The anonymous article began with the observation that according to Ministry of Justice figures, the German sterilization law was being "widely applied," with more than 56,000 operations in 1934. Hence, the author thought it wise for Catholic moral theologians to review the issues and questions arising from the German sterilization plan. The article posed three key questions: "(1) What is the relation of the individual to the State?; (2) What is the true understanding of the term 'common good' to which those who propose state-sterilization of the unfit appeal?; (3) How does the true understanding of the term 'common good' apply to the question of state-sterilization?" From the phrasing of the questions, it is immediately apparent that the author considered the Nazi theories of the state and community to be false. To address the proposed questions, the author applied the principles of Catholic moral theology to the individual elements of the German sterilization law, whose entire text was reproduced in the *Review* as an appendix. In concluding his reflections, the author rejected the German laws, "because they deify the State and sacrifice all individual freedom and the most sacred natural rights to a falsely conceived common good."[46] Thus he reiterated the stance of Catholic moral theologians regarding sterilization that had been crystallized with the release of *Casti Connubii*: the state has no control over individual bodies, and legislation motivated by base self-interest undermines the true purpose of community.

The US Catholic press thus used German developments to reinforce the teachings articulated in *Casti Connubii*. This task was made easier by Pope Pius XI's repeated condemnations of German racialism. US Catholic publications could point directly to the words of the holy father when they urged their co-religionists to fight eugenic sterilization measures at home in the same manner as those Catholics around the world who were also struggling against eugenics efforts.

Ambivalent Territory

Reflecting on developments in Germany was not the only way that Catholics in the United States meditated on the proper relationship between individual rights, civil power, and racial hierarchy. For example, the editors of *Commonweal* noted in April 1935 that eugenics and sterilization held a wide appeal across Europe and the United States. With the possibility of war looming, police in Bordeaux, France, reported that they had discovered a group who hoped to avoid military service by getting sterilized. At the instigation of the French, according to the Associated Press, the authorities in Brussels had arrested a German communist accused of encouraging the group in Bordeaux. Relating the tales of these developments in western Europe to their US readers allowed the *Commonweal* editors to seamlessly shift to a survey of legislative activity surrounding sterilization in the United States, including activity in Georgia, Iowa, Missouri, New Jersey, and Pennsylvania. The editors also noted the troubling and significant possibility of coercion and abuse that accompanied eugenic sterilization policies, pointing out that "charges that men and women had been forced by relief workers, under penalty of being denied relief, to submit to sterilization operations in Contra Costa County, California, were being investigated, while a similar charge was made in Ohio."[47] Thus, reflections on the international status of eugenic sterilization easily lead to a reflection on the local status of eugenics issues.

During the mid-1930s, alongside the emergence of Nazi propaganda in Germany, a push for eugenics legislation closer to home in Puerto Rico attracted the attention of US Catholics. In its peculiar status as a US territory under which its citizens were unable to participate in national politics, Puerto Rico posed an interesting problem for US Catholics. While primarily Catholic, the Puerto Rican population thought very differently about American political traditions and social mores than did their mainland counterparts. Hence, in discussing eugenics developments on the island, mainland Catholics tended to treat Puerto Ricans as citizens of a discrete nation rather than as fellow Americans. With the distance created by this perspective, Catholics on the mainland used their reflections on the Puerto Rican situation as a cautionary tale about the "fallacious short-cuts" of modern social reform.

In informing their readers about eugenics developments in Puerto Rico, the Catholic press did not adequately represent the tumultuous struggles for social and political power within Puerto Rico that were parallel to and tied up with efforts to pass and implement legislation dealing with reproductive issues. Rather, the editors of Catholic periodicals used the news of contraceptive, marriage, and sterilization legislation as a vehicle to reiterate the Vatican's teachings from *Casti Connubii* and to speculate that the legislative status of compulsory sterilization in the United States might change for the worse at any moment. In May 1934, the editors of *Catholic Action* informed their readers that Bishop Edwin V. Byrne (1891–1963) of San Juan had responded to the introduction of eugenics legislation with a strongly worded statement that repeated almost word for word the text of the encyclical as it related to eugenics and sterilization. Significantly, Byrne had prefaced the statement by explaining that the teachings on eugenics and sterilization applied "to all men, regardless of party or creed, and their intrinsic value affects savage or backward countries as well as the most cultured and progressive communities." By framing his remarks in this way, Byrne clearly indicated that he did not consider the Puerto Rican community to be "cultured and progressive." Nonetheless, he counseled the people of Puerto Rico that the right to marriage and to have children was a natural right for each person, one that the state could not justifiably usurp. But, he continued, while the state has no power over individual bodies, neither do individuals have the right to "destroy or mutilate their own members except in the case where there is no other way of assuring the welfare of the body as a whole." In addition to restating the position of the Vatican, Byrne quoted extensively from the work of Jacob H. Landman, a Jewish attorney who advocated sterilization in certain cases but who argued that many of the conditions for which sterilization was a presumed remedy were not hereditarily transmissible. Based on those moral and scientific grounds, Byrne reminded the Puerto Rican community of their duty as a mostly Catholic citizenry to resist the passage of eugenics legislation.[48]

While using Bishop Byrne's statement to repeat the teachings of Pope Pius XI's encyclical on marriage, the editors of *Catholic Action* did not pass up the opportunity to point out to their readers that as citizens of the United States, they faced the same threat from eugenics legislation as did their co-religionists in "savage" and "backward" Puerto Rico. The editors portrayed Puerto Ricans as an endangered community that was being manipulated and abused by eugenics advocates and arrogant reformers. Catholics in the United States might be better equipped to fight such policies. At the end of the piece, the editors pointed to the four pamphlets produced by the National Catholic Welfare Conference as useful for discussion of the "question of eugenics and the best scientific and medical knowledge on the value of sterilization." Touted as pamphlets "written by experts in their respective fields," the entire series could be obtained for only 25 cents. Thus, the editors of *Catholic Action* took the events in Puerto Rico as an opportunity to educate their readers on issues associated with eugenics and sterilization, hopefully increasing the ranks of Catholic citizens who would actively oppose similar legislation on the mainland.

Despite the urgings of Bishop Byrne, the eugenics legislation introduced in 1934 was eventually passed in the Puerto Rican legislature, and was signed into law in 1937. The three laws provided for marriage restrictions, set up a eugenics review board for compulsory sterilizations, and made contraception legal. In response, Bishop Byrne issued a pastoral letter to the people of Puerto Rico that urged them to lobby their legislators to repeal the measures, based on the Vatican's teaching that the measures were contrary to natural law.[49]

Although Bishop Byrne's statements do not reflect it, advocates for eugenic sterilization took advantage of a long-standing cultural tendency to link poverty and overpopulation. Mainland reformers called for solutions to the overpopulation problem rather than for a redistribution of power and wealth. Eugenicists played a role in this conversation by advocating programs to curb the reproduction of the working classes on the island. In doing so, they echoed the long line of eugenicists, such as Madison Grant and Lothrop Stoddard, who had sounded the alarm about differential birthrates among peoples of color around the world. From this perspective, mainland conservatives called for a continuing role for the United States in helping to "fix" Puerto Rico's overpopulation.[50]

Writing for the Jesuit journal *America*, Barrett McGurn reported on developments in Puerto Rico and used that information to criticize birth control and sterilization as ineffective solutions for what he perceived to be a problem of economic resources, not overpopulation. In the face of unsuccessful New Deal reforms, Rafael Menendez Ramos, Puerto Rico's acting governor and a Penn State University–educated *Birth Control Review* subscriber, began to voice his support for legalizing contraceptive devices and instruction, explaining: "It looks innocent, but I hope to see birth-control clinics all over the island set up and run by the insular government." Additionally, the legislature passed marriage restrictions designed to prevent the union of persons with certain mental and physical diseases, and provided for a board to authorize the sterilization of the mentally retarded. Most significantly, however, unlike any of the states on the mainland, the Puerto Rican legislation provided not only for eugenic sterilization but also for voluntary contraceptive sterilization.[51] To McGurn, this corpus of legislation was a bald-faced attempt by the Puerto Rican government to control the reproductive activity of the island's population. He expressed his surprise and dismay that in a predominantly Catholic community officials could exhibit "ideas of religion and of morals" that were "so twisted, so typically pagan, that a conversation with them can only depress."[52] McGurn suggested that, in the minds of persons like Acting Governor Ramos, common mores were being construed as morals and were replacing moral principles with convenient answers to complex ethical problems. Rejecting these developments in contraceptive and sterilization policy, McGurn explained to his readers that Puerto Rican peasants should be convinced to "migrate to those corners of the world where a flagging birthrate has brought forth premiums for babies and penalties for bachelors." McGurn also urged Puerto Rican elites to work to convince the US Congress to ease regulations and import duties on sugar and to attract investors for other island industries, such as needlework, tobacco, rum distilling, fruit packing, and candy making. In this argument, McGurn took the path offered frequently by other Catholic thinkers, such as John Ryan—one that rejected reproductive restrictions in favor of economic justice.[53]

Thus, commentary about eugenics legislation in Puerto Rico served to reflect the possible consequences of not offering strenuous opposition to eugenics policies and of not advocating for economic justice. Bishop Byrne repeated the sentiments from *Casti Connubii* and the Catholic journals promoted NCWC literature on eugenics sterilization as a way of reminding their readers of their duty to speak out. By focusing on the predominantly Catholic population of Puerto Rico, the US Catholic press was issuing a sharp warning to its readers that if they lost sight of their values and moral obligations, then they too could find themselves in a situation where economic problems were being solved with surgical solutions. With the specter of the Depression close at hand, US economic conditions, while more robust than those in Puerto Rico, were still fragile enough to cause alarm. So, while the US Catholic press failed to fully reflect and explore the deep social and political complexities that formed the context for the Puerto Rican eugenic legislation, they did succeed in offering an object lesson for their readers on their social and political obligations.

As the Puerto Rican example makes clear, US Catholics who were active in their opposition to eugenics policies found events abroad useful in their efforts to articulate a response to the eugenics movement at home. Whether they were reflecting on German sterilization laws, on the condemnation of sterilization by Pius XI, or on poverty in Puerto Rico, Catholic activists used those situations to instruct US Catholics on the official teaching of the Roman Catholic Church with respect to eugenics. Fortified by the publication of *Casti Connubii*, US Catholic opponents of eugenics felt that they had the support of the highest moral authority when they spoke out about sterilization. This claim to ultimate authority was in sharp contrast to the claims of scientific authority that the eugenics movement used to justify their public policy initiatives.

One key to *Casti Connubii*'s message was the perspective of Catholic teaching on the relationship of the individual to the state. The condemnation of eugenic sterilization articulated in the encyclical proscribed the authority of the states, so that they could not violate the individual's ability to marry and raise a family simply on the grounds that the offspring might not measure up to a eugenic standard of fitness. In making this prohibition, the Vatican elevated Catholic moral teaching above both the power of the state and the authority of science. While scientific and legal arguments would still be important for Catholics as they engaged with one another in the public sphere, the moral teaching about the inviolability of an individual's human rights and bodily integrity offered welcome clarity and support in their battle against negative eugenics policies. The encyclical made clear that the state could not exercise the ultimate control over the lives of its citizens and that science was not the arbiter of human value. By focusing on the integrity of the individual, Catholics in the United States circumvented popular claims for reform based on the "common good" and scientific authority. At the same time, some non-Catholics questioned the willingness of the magisterium to offer pronouncements on marriage and family that were to apply to all persons, not just Catholics. To some extent, this revived concerns about Catholic thinking on the relationship between the church and the state.

Members of the Catholic press looked with alarm upon developments in eugenics around the world. They saw the broad encroachments upon human rights that existed in Germany, Puerto Rico, and other areas as possibly presenting a similar risk within the United States. Catholic activists worked to remind their co-religionists of the scientific dispute that surrounded eugenics and of their obligation to oppose eugenics policies on a local level. So, even as eugenics advocates in the United States hailed the news of sterilization legislation in other nations as an affirmation of the growing acceptance of their scientific authority, Catholics drew affirmation from the pope's teaching and remained vigilant in their opposition to such measures at home.

FIVE

The Greatest Obstacle: The Growth of a Confident Opposition

Even when discussing the growth of eugenics in other national contexts, authors in Catholic journals and periodicals almost invariably returned to their own backyard. They frequently pointed to the dangerous situation in state legislatures, where sterilization statutes and marriage restrictions kept returning to the agenda. Based on the success of the Virginia sterilization statute in 1927, eugenics advocates around the country pushed for similar legislation throughout the 1930s and well into the 1940s. A curious contradiction occurred in the years before and during World War II: amid the growing scientific criticism of sterilization there was a simultaneous growth in the number of sterilizations performed.

As the 1930s progressed, an increasing number of medical research groups and scientific organizations rejected eugenic sterilization as a realistic and efficacious way to deal with mental deficiency. This growing coalition of critics bolstered Catholics' confidence in their stance against eugenic sterilization laws. As a result, during the 1930s Catholic intellectuals and the Catholic press produced a significant amount of literature on sterilization. At first this literature was directed primarily at their co-religionists, as had been the case for the previous fifteen years, during which time the articles that appeared in Catholic journals and newspapers and the National Catholic Welfare Conference pamphlets had sought to articulate the moral teaching on eugenics. Catholic writers in the 1930s continued to embrace those non-Catholics who opposed sterilization, and they sought to remind Catholics of their duty to be vigilant in their opposition to such measures. But as the decade wore on, Catholic writers began to more explicitly address a larger, non-Catholic audience.

The impact of this new approach was not lost on eugenics supporters. For example, when Marion S. Olden, a vocal sterilization advocate from

New Jersey, reviewed the state of the nation's sterilization programs in 1935, she discovered that there was "but one source of all the active opposition that exists in the whole United States—the Roman Catholic Church."[1] Olden's rabid anti-Catholicism might have led her to exaggerate the singularity of the Catholic opposition, but she was not wrong to point out that Catholics were consistent and vocal opponents of sterilization legislation in many states around the country, especially after *Casti Connubii*. By the middle of the 1930s, Catholic intellectuals and writers were so confident in their opposition to negative eugenic policy initiatives that they increasingly engaged with eugenics advocates in debates that reached secular and, often, non-Catholic audiences.

The flurry of activity surrounding eugenics policy initiatives in the 1930s took place during the cultural and economic cataclysm of the Great Depression, and was significantly influenced by it. Given their long-standing role as purveyors of charity and social provision, Catholics had an understanding of the social milieux for which sterilization was not a viable solution to economic hardship. In these years of great hardship, Catholic activists worked to achieve a seemingly impossible balance: urging the expansion of relief measures from the state, and protecting individuals from state intrusion into marriage and family life. As the nation seemed to veer toward collapse, eugenics supporters pushed a long-standing but suddenly more prescient rationale for sterilization: it represented one way to ease the burdens that delinquents and defectives placed on the resources of the state. To Catholics, such arguments remained abhorrent, no matter the circumstances.

Objecting in Ohio

Throughout the 1930s, as they had for the previous decade, Catholics worked effectively in the individual states to thwart eugenic sterilization statutes. Just as in the past, this work often involved a combination of ecclesiastical pressure and lay activism. But within these established strategies we can also witness a growing willingness of Catholics to assert their point of view and their political leverage. In the effort to influence state legislators, members of the church hierarchy and dedicated laypersons waged a multifaceted campaign that demonstrated the degree to which Catholics now saw themselves as fully invested in American political institutions. They lobbied individual legislators, the press, and key party officials, all in the hopes of influencing the legislative process. In doing so, they shrewdly worked the political system by asserting the power of the Catholic vote and by making use of a network of sympathetic politicians and welfare providers to press their cause.

Once again, the Ohio historical record provides a particularly useful look at this activity. In January 1931, and then again in 1933, the Ohio state legislature took up the issue of eugenic sterilization, just has it had in 1927 and in 1929. However, in the 1930s the bishop from Cleveland offered a far more extensive and vigorous opposition campaign. Bishop Joseph Schrembs, acting in consultation with Bishop James J. Hartley of Columbus, drew on a variety of resources in orchestrating his protest against the measure. He turned to nationally noted Catholic thinkers to develop the key points of his response. Then he arranged for his representatives to testify before the legislature and wrote letters to legislators and the local media registering his opposition to the law. Finally, Schrembs lobbied behind the scenes to secure assurances from individual legislators that they would oppose the sterilization measures.

Schrembs used the writings of John Montgomery Cooper to formulate his response to the 1931 bill, composing a letter from a brief that Cooper had authored in 1929. At the end of January, Schrembs sent his letter protesting the Emmons Sterilization Bill to all of the representatives and senators in the Ohio State legislature.[2] The letter rejected the four common grounds usually submitted as reasons for sterilization legislation: (1) that crime among the feebleminded was increasing; (2) that the so-called feebleminded had an unusually high birthrate; (3) that the mentally deficient were disproportionately increasing as a percentage of the population; and (4) that efforts to educate feebleminded children were wasted. Schrembs pointed out that there was no reliable statistical evidence to support any of these claims. He commented, for example, that some programs demonstrated that careful pedagogy among the "feeble-minded" could produce active, self-supporting members of the community. The letter also argued that forced sterilization was a grave transgression of fundamental personal liberties, and moreover that it was an ineffective solution to the problems associated with feeblemindedness, since recent scientific evidence suggested that at least 50 percent of mental deficiency was not hereditary in nature.[3]

Bishop Schrembs received considerable, and varied, responses from the members of the Ohio legislature. One representative simply pledged that the bill would be given "due consideration" and that Schrembs's arguments would receive "proper recognition," but others responded in more detail.[4] Senator Nils P. Johnson of Youngstown, for example, explained that the Emmons measure raised matters that were "highly technical" and that would "require expert opinion and advice" before he would be "at all qualified to vote intelligently on the bill." Johnson continued: "Regardless of the position which we might take on this question of sterilization, all of us must give some thought to the work and progress that is being accomplished in the lower animal kingdom, especially in the breeding of stock and high bred horses. Then again many instances can be given where genius has been produced from the lowliest and comparatively inferior parents." Given these contradictory indicators, the senator suggested that he would be guided by the administrators of the state's institutions who had the most experience serving the best interests of the mentally ill. In the end, Johnson thanked Schrembs for his input but did not indicate how he would vote on the bill.[5] Robert T. Secrest, on the other hand, found himself quite moved by Schrembs's letter. Since he usually did not make up his mind on bills before the floor debate, he claimed never to have answered a lobbyist's correspondence. However, he made an exception for the bishop because he found the letter to be "the most gentlemanly protest against any bill that I have ever observed." Secrest indicated that at that point his "attitude toward the bill is one of disapproval."[6]

A number of other senators and representatives indicated their solidarity with the bishop and offered varying levels of assistance in defeating the sterilization bill. Still others pledged that they would be sensitive to the concerns of the church. Representative Lawrence Welsh, of Steubenville, explained that in the previous two weeks he had been "endeavoring to line up enough votes in the House to defeat" the bill. He added, "It behooves those of us who are aware of the full content and intent of the Bill to labor unceasingly that our efforts will succeed." Likewise, Cleveland representative A. S. Harding claimed to have had contact with many people who promised to work for the bill's defeat. Senator Joseph N. Ackeman, also of Cleveland, had not only studied the contents of the bill but he had also consulted with a local priest in forming his response to the legislation. Ackeman promised Schrembs: "I shall do everything in my power to prevent the enactment of the . . . legislation, and I shall appear before the proper committees and fight it on the Senate floor with all the energy that I possess." Ackeman also assured Schrembs that on matters coming before the Ohio legislature, "I shall be only too happy to receive your counsel and advice, as I know that any advice I may receive from you will be for the benefit of humanity."[7] Together, these responses show that, for at least some legislators, Schrembs held an esteemed position on moral issues.

On the other hand, Representative James C. Foster was concerned that the bishop did not properly recognize Foster's own role as a "watch dog" for the church on matters coming before the legislature. Foster explained that his wife had attended a meeting of a woman's group in which another representative had been given credit for leading the opposition on the sterilization bill. Troubled by this oversight, Foster exhorted the diocesan chancellor, Monsignor James A. McFadden: "You will recall that I told you personally on my visit to the Bishop that it was my intention to be the watch dog in the Lower House, and what I am endeavoring to do is to build up a future in political life in Cleveland, and I believe that the fundamentals of anyone's political life is based upon the services he has rendered to the constituents, providing they give him credit therefor." Foster went on to emphasize the political cost of choosing to champion the Catholic perspective on issues that came before the legislature. Foster was correct that his support was important to the Catholic cause, since Ohio had the sixth-highest concentration of Catholics in the country—a population that hovered around 1,000,000 in 1930.[8] Foster's complaint is a demonstration of the power that Schrembs wielded among Catholic constituents of the Ohio legislature.

In the middle of February, Schrembs received a note from Sara Varley, the future religion editor for the *Cleveland Plain Dealer* and an active member of the diocesan Council of Catholic Women, reporting that the head of the Youngstown Community Chest was actively promoting sterilization legislation.[9] This notice was particularly important since the Community Chest was the city's cooperative social relief fund, distributing resources to many charities in the area, including Cleveland's Catholic charities. While Catholic charity organizations sometimes accepted such federated funds, they worked hard throughout the Depression to sustain their own central role in providing welfare relief for their co-religionists. By 1932, the Depression's effects were so widespread that 80 percent of relief came from public sources. The National Council of Catholic Charities lobbied the state and federal governments for the right to participate in the distribution of social provision to clients based upon the principle of "subsidiarity" as articulated in *Quadragesimo Anno*, Pope Pius XI's 1931 encyclical marking the fortieth anniversary of *Rerum Novarum*. "Subsidiarity" meant that local Catholic charities should cooperate with larger social organizations to provide services to the needy, which would allow Catholics to care for the spiritual and material needs of their own. Initially, between 1929 and 1932, local Catholic charities made an effort to meet the needs of their clients on their own. But finding that their resources were inadequate, they later applied first for state funds and then for federal moneys. The pattern repeated itself in New York, Pittsburgh, Cincinnati, Denver, Baltimore, and other cities around the country. Thus, as the programs of the New Deal developed and relief filtered down through the public system, Catholics maintained a significant structure for social provisions.[10]

Sara Varley's warning that the leader of the Youngstown Community Chest was supporting sterilization meant two things for Schrembs. First, it suggested that Catholic social workers and the clergy would have a significant moral conflict with the agencies with which they interacted on a regular basis. Second, it suggested that relief officials in Ohio were embracing sterilization as part of an effort to deal with the economic burdens of the Depression. In response to Varley's note, Schrembs addressed an extensive letter to both the *Youngstown Vindicator* and the *Youngstown Telegram*, in which he argued that medical and scientific knowledge contradicted the claims of efficacy by advocates of sterilization for the mentally ill. He cited the work of Dr. H. H. McClellan, a non-Catholic who was the former superintendent of the Dayton State Hospital for the Insane. Dr. McClellan had argued that family studies of the mentally ill suggested that their families contained no higher incidence of mental illness than those of persons who were not mentally ill. In closing, Schrembs emphasized the responsibility of politicians to protect individual rights even in the face of the hardship of the Depression. He explained, "In the crowding cares of the twentieth century there is a marked tendency to consult the community welfare at the expense of the individual. Yet there are certain personal liberties that may not be curtailed, save for the commission of crime. One of these, certainly, is life and the integrity of the human body." Here, Schrembs was highlighting the important balance between government aid during the Depression and safeguarding individual rights. Given the centrality of these rights, and the medical opinion rejecting sterilization, Schrembs counseled his readers: "It is well in these matters to follow the advanced thought of the real leaders of the medical profession rather than the chicanery of politicians."[11]

In his effort to defeat the Emmons bill, Bishop Schrembs found himself embroiled in the tangle of Ohio politics. In addition to directly lobbying the representatives and senators by letter, he attempted to influence key political players in the Republican Party, which was championing the Emmons bill. First, he sent two Catholics from Canton to meet with Senator Rufus Finefrock, the chairman of that body's Sterilization Committee. They encourage the senator to stop the bill in committee, but Finefrock informed them that the question was a political one that would require him to stand by his party. Undaunted, Schrembs decided to pay a visit to Maurice Maschke, the Republican Party boss. Maschke told him that the bill would not pass. However, shortly thereafter, Maschke called to ask if

Schrembs could persuade Governor White to oppose the bill because he was unable to prevent the Republican legislators from voting the bill out of committee. According to Schrembs, "The whole thing seems now to have come to the stage of a political game where the Republicans are willing to trade the Sterilization Bill for some consideration from the Governor." Schrembs then asked the bishop of Columbus, James J. Hartley, to approach White.[12]

Schrembs continued to track the progress of the bill through committee well into March. On March 10, in a final effort to influence legislators, he sent a telegram to five state senators in which he succinctly stated his objections to the Emmons bill: "Sterilization Bill as well as Amendments are basically wrong and objectionable. Sterilization is an unwarranted invasion of fundamental human rights. It opens highway of promiscuous licentiousness and will promote spread of venereal diseases. I protest against this iniquitous invasion of fundamental human rights."[13] By raising the specter of unrestrained promiscuity, Schrembs invoked the fear that the modern world was descending into chaos and immorality. This stance is more than a little bit ironic given that administrators at state institutions used the standards of respectable behavior and order to condemn their charges to compulsory sterilization. Regardless, the bishop's unrelenting lobbying paid off; the Emmons bill failed in the spring of 1931.

Despite this defeat, Ohio sterilization advocates were undeterred. Early in 1933 they again brought a eugenic sterilization measure before the legislature. This time, Bishop Schrembs enlisted Cleveland's active Catholic women in an attempt to persuade the legislators to vote the bill down in committee. The president of the Cleveland Diocesan Council of the NCCW, Lillian M. Westropp, and the legislative chairman, Alice F. Kelly, registered their protest against the 1933 Emmons bills with senators from Cuyahoga, Akron, and Youngstown. Their letter informed the senators that the Diocesan Council, representing women's organizations in fourteen northeastern Ohio counties with an approximate membership of twenty thousand, had passed a resolution denouncing the law.[14] To supplement this message from the council, Bishop Schrembs sent a telegram to the members of the state senate's Public Health Committee, in which he repeated almost word for word the objections that he had made in 1931.[15]

In addition to registering this official protest with the senators from the area, Alice Kelly did some checking behind the scenes. First, she ascertained the occupation and religious affiliation of as many of the members of the Public Health Committee as she could. According to her research, the committee contained two Catholics, one Protestant, and four senators whose faith was unknown. Second, she consulted John Smolks, a Catholic senator from Cuyahoga County with whom she had a personal friendship. Smolks informed her that three persons on the committee were definitely opposed to the sterilization bill, so in order to ensure defeat, they only would have to sway the opinion of one other member. Kelly reported to Bishop Schrembs that Smolks and the other senators opposed to the bill had suggested that "a little pressure should be brought on Mr. W. H. Annat." As an attorney from Cuyahoga County, Annat was also a member of the same law firm as the Catholic "welfare relief man," A. V. Cannon. Kelly suggested that Schrembs contact the chairman of the local Democratic Party, who should in turn speak with Annat. Additionally, she thought passing a word to Cannon would also help.[16] Thus, Kelly's background work revealed a way that they could use the existing network of Catholic charity workers to bring some influence to bear on the situation.

Though there is no existing evidence that Bishop Schrembs spoke with the head of the Democratic Party or with Cannon, it is likely that he found some way to make contact with Senator Annat because on February 2, 1933, Schrembs sent a note to Annat thanking him for the news that the sterilization bill had been voted down in committee.[17] Schrembs's consistent and politically shrewd efforts to defeat the Emmons sterilization legislation in Ohio helped build the impression among eugenics advocates that Catholics were becoming a major obstacle in their attempt to pass sterilization legislation across the country.

Sterilization in Hard Times

The introduction of sterilization legislation in Ohio was in part precipitated by the Depression. This was true of increased eugenics activity in other states as well. The Supreme Court's affirmation in 1927 of the state's right to sterilize an institutionalized patient without that person's consent, followed by the collapse of the stock market in 1929, created a unique opening for eugenicists. First, a concept that already tickled the racial imagination and desires for social engineering in a broad swath of Americans then found affirmation from the highest court in the land; and then the nation experienced the most devastating economic collapse in its memory. The results were immediate. The number of states with involuntary sterilization statutes increased from nineteen in 1927 to twenty-seven by 1932. This increase coincided with the dramatic drain on state resources created by the Depression.

The historian Philip Reilly analyzed sterilization activity during the period from 1907 to the beginning of World War II, and he noted a significant spike in sterilization during the Depression. In the fourteen years between 1907, when Indiana's sterilization statute was passed, and 1921, at least 3,233 institutionalized persons were sterilized in the United States, whereas in the fourteen years between 1927 and 1941, nearly 30,000 sterilizations were performed. State sterilization programs were more active during the 1930s than in any other decade. The most pronounced jump in the number of operations began in 1929. Whereas in 1925 only 322 sterilizations were performed, in 1929, 2,362 people in the United States were sterilized. By 1932, that number had jumped to 3,921. From 1933 until 1941, yearly totals hovered between 2,000 and 3,000 sterilizations, resulting in a cumulative total of 38,087 sterilizations for the Depression years.[18] This suggests that eugenic sterilization remained popular even, as we shall see, in the face of growing doubt in the scientific community about its efficacy for advancing eugenic goals.

Reilly goes on to argue that with the onset of the Depression, officials in charge of ordering sterilizations "became less concerned with preventing the birth of children with genetic defects and more concerned with preventing parenthood in those individuals who were thought to be unable to care for children." Sterilization and parole were ordered for less severely impaired persons to make room for more patients and to ease the economic burden on the state. This strategy is evidenced in the other crucial shift—in the sex of those sterilized—that occurred during the Depression. Around 1930, the number of women sterilized began to increase. Up through the end of 1927, the proportion of institutionalized persons sterilized was roughly 54 percent male to 46 percent female. Then, between 1928 and 1932, that proportion shifted, with fully 67 percent of sterilizations performed on women. Through the remainder of the 1930s, the population of institutionalized persons subject to sterilization maintained a rough ratio of 40 percent men to 60 percent women. Reilly argues that this shift in the number of young women who were sterilized was related to economic concerns and parole issues: a paroled feebleminded woman was far more likely to reproduce than a feebleminded man. Thus the Depression context influenced the use of sterilization in state institutions, resulting in an increase in procedures predominantly affecting women.[19]

Throughout the 1930s the Catholic press noted the increase in sterilization activity and advocacy and reported on sterilization legislation around the country, exhorting readers to be aware of changing conditions and of the need for a vocal opposition. This flow of information to the Catholic reading public was facilitated by the NCWC News Service. Under the guidance of Patrick Ward and Frank Hall, the service provided diocesan newspapers and national publications with bulletins detailing the legislative activities of eugenics advocates and the responses of local Catholics to those activities. In many cases, those news bulletins were simply reprinted word for word in newspapers and journals, but on other occasions, they prompted extensive commentary from local editors concerning the eugenics movement, its negative policy initiatives, and the duty of Catholics to resist its legislative efforts. Thus, the readers of Catholic publications were persistently reminded that the teachings of their faith placed them in conflict with the advocates of a popular and potentially dangerous social movement.

In March 1931, the editors of *America* used the occasion of a Nebraska Supreme Court ruling to remind Catholics of the church's position on eugenic sterilization and to exhort them to play an active role in fighting any corresponding legislation. In the ruling, the Nebraska Supreme Court upheld that state's 1929 law, which called for the sterilization of feebleminded persons as a prerequisite for release from state institutions. In response, *America*'s editors quoted extensively from those sections of *Casti Connubii* that addressed the danger of placing the cause of eugenics and the power of civil authority above the state's obligation to protect the integrity of human persons. The editors also sought to bring the argument to non-Catholic readers by pointing to the National Committee for Mental Hygiene, which had declared that sterilization statutes were based on dubious assumptions about heredity. They also invoked the words of the governor of Pennsylvania when he had vetoed a 1905 sterilization statute. "One wonders just where the *vox populi* may be when legislation of this sort is being discussed," they mused, "and why especially Catholic men and women seem so apathetic." Exhorting their readers to action, they explained, "Many years ago, Leo XIII laid down in no uncertain terms that one of the chief Christian duties of a citizen is to bring his influence to bear to ward off from his fellow-citizens legislation which contravenes the laws of nature or of God."[20]

Despite the editors' perceptions of Catholic apathy, reports in *Central-Blatt and Social Justice,* the journal published by the Central Bureau of the Catholic Central Verein of America, a German American Catholic fraternal organization based in St. Louis, suggested otherwise. In one issue of the journal, it was reported that active opposition to sterilization statutes was taking place in at least four states. In February and March of 1931, the bureau circulated an information packet to priests and laypersons in Texas, Ohio, Missouri, and Indiana that was designed to assist in the efforts to combat sterilization statutes. The packet included copies of the bureau's leaflet "The Case against Sterilization of Mental Defectives," which discussed scientific issues and the responsibilities of institutions to care for the welfare of their patients.[21] In Texas and Missouri, members of the local Central Verein Committees on Legislation distributed copies of the leaflet to members of affiliated societies as well as to members of the state legislatures who might be disposed to reject sterilization laws.[22]

The Catholic press also took note of the rapidly increasing numbers of sterilizations taking place. For example, *America* noted with alarm that according to the American Legislators Association, in October 1934, more than 16,000 persons had been sterilized in the United States. In the face of these figures, the editors reminded their readers that there was "no guarantee—rather the contrary—from science that sterilization achieves its supposed effects."[23] Similarly, in 1935, *Commonweal* published the results of a United Press survey showing that nearly 20,000 persons had been sterilized under the eugenic regulations of eighteen states. Of those persons, roughly 5 percent were judged criminals, 50 percent were labeled "insane," and 40 percent were "feeble-minded."[24] By publicizing figures on the frequency and the rationale for sterilizations, Catholic periodicals worked to instill a sense of urgency in their readers. The public was ignoring scientific evidence in supporting sterilization, and without unrelenting Catholic opposition, the dramatic increase in the number of operations would continue.

The sudden economic appeal of sterilization starting in 1929 was mirrored over the next decade by significant developments in the sciences. The Catholic press, much to its delight, was now able to point to a growing number of non-Catholic physicians, biologists, anthropologists, and social theorists who shared their objections to eugenics policies, practices, and ideologies. With the swell of scientific doubt, it appeared that the tide was beginning to turn toward the position that had always been expressed by Catholic intellectuals and activists. Both *America* and *Commonweal* took special note of an article written by Dr. Morris Fishbein that appeared in the March 3, 1935, edition of the *New York Times*. Fishbein was the editor of the *Journal of the American Medical Association*. In his *Times* piece, by citing statistics about the actual intelligence of parents and children subjected to sterilization—some could be characterized as "normal," some "super-normal," and some "sub-normal"—Fishbein had criticized eugenic sterilization as an ineffective way to deal with hereditary degeneracy. In its report of Fishbein's position to their readers, the editors of *Commonweal* explained, "In the face of the growing sentiment for sterilization as a social cure-all, it does no harm to remind ourselves how divided medical opinion is on the practice, and how many data from disinterested sources can be mustered against it."[25] A similar article appeared in *America*.[26] Likewise, the findings of the American Neurological Association's Committee on Eugenic Sterilization cheered the editors of Catholic journals. Under the direction of Dr. Abraham Myerson, a psychiatrist, the committee affirmed the 1934 findings of the Brock Committee. In a study sponsored by the British government, the Brock Committee had concluded that the understanding of the laws of heredity was not sufficiently developed as to justify the sterilization of the "unfit." Paul Blakely was happy to summarize some of Myerson's statements to the *New York Times* regarding heredity, the unfit, family size, mental disease, and crime, for the readers of *America* in 1936. He seconded Myerson's call for further research, explaining: "There is, then, no justification in science for sterilization voluntarily permitted by the patient or his parents. Sterilization made compulsory by the state is, of course, pure tyranny."[27]

The notion among Catholic writers that the tyranny of state-sponsored violence was widespread was created by the sheer numbers of sterilizations, and reports of grave abuses in sterilization operations performed by local institutions only amplified that perception. Perhaps the most egregious instance of such abuse came to light in Kansas in 1937. In October of that year, Kathryn O'Laughlin McCarthy, a former member of Congress and a Catholic, called for an investigation into the sterilization practices at the Beloit Industrial School for Girls. McCarthy had been approached by the superintendent of that institution about the fact that the school needed new rugs but was low on funds, due to having spent $4,000 on sterilization operations and hospitalization costs. After further investigation, McCarthy discovered that 62 of the 148 girls at the institution had been sterilized, and that 22 more were scheduled for the operation prior to an administrative change. According to McCarthy, some of the parents of the young women had objected to the operation but were given insufficient notice to wage any protest. McCarthy questioned the clinical justification for the procedures. "In one instance I found it was because a girl was high tempered," she explained. "So far as I can determine, it was done as a punishment, rather than for any special good of society." The administrators of the state correctional school maintained that the girls and their parents were given hearings before the school's superintendent, physician, and a board of administration, and that the system had adhered to the Kansas statute. Despite these protestations, the new administrator at the school, Blanche Peterson, confirmed that under the previous administration, the girls, who were all under the age of sixteen, had lived in fear of being sterilized, and that the operations had been done without proper consent and for insufficient reason.[28]

For Catholics and non-Catholics alike, this detailed and disturbing report contributed to the hostility toward compulsory sterilizations. But for Catholics especially, who had repeatedly heard clergy and lay activists caution that eugenic sterilization statutes were ripe for abuse, McCarthy's report was a confirmation of nearly three decades of grim predictions. In combination with the reports of the increasing numbers of sterilizations, stories like McCarthy's created the impression of potentially widespread human rights abuses in the nation's industrial schools and homes for the feebleminded. Even the very minimal safeguards of human rights that were built into sterilization legislation were being ignored, and young women were being forever robbed of their ability to bear children. These abuses occurred despite the slow but growing scientific criticism of negative eugenic policies. But in the dire circumstances of the Depression, the powers of the state and of individual administrators took on new potency and posed an increasing danger for the inmates in their charge.

Diving into the Secular Press

Encouraged by the findings of medical research groups and scientific organizations headed by non-Catholics, in the mid-1930s Catholic activists grew more assertive in their condemnation of eugenic sterilization. Catholic writers' calls for activism were now made with the hope that right would triumph over the misguided eugenics movement. No longer did they feel alone, comforted only by the moral teachings of their faith; now, they were emboldened by the growing weight of scientific opinion. Popular discussions of sterilization and eugenics in the secular press gave Catholics an opportunity to bring their long-established positions to a much wider and more diverse audience. When Germany implemented its national sterilization laws in 1934, Catholic intellectuals were invited to express their views on eugenics in *Scientific American* and in the *Forum*. By that point, it was clear to most persons informed on eugenics issues that Catholics in America had mounted the most fully developed and organized opposition to sterilization and they were more than willing to share their critique of the movement with the broader American public.

In the summer and fall of 1934, the well-established and popular science magazine *Scientific American* carried a four-part debate about sterilization. Significantly, the editors sought out contributions from a range of voices on the issue, clearly suggesting that sterilization was not universally accepted as a remedy for what ailed the human community. The first article in the series was written by Jacob Landman, a Jewish attorney who supported sterilization in select instances but called for greater scientific investigation and caution in the use of the procedure. The second contribution was from Ezra Gosney, head of the Human Betterment Foundation and an enthusiastic supporter of the widespread use of sterilization in public policy and social reform. The third piece, by the German physician C. Thomalla, was secured from a spokesperson for the German embassy and represented the official position of that government, reiterating the principle that economic savings justified the state's intervention in the reproductive lives of defectives. The final contribution, which communicated the Catholic opposition to the eugenic sterilization movement, was from Ignatius W. Cox, a Jesuit ethics professor at Fordham University who had strong ties to Catholic physicians' organizations.

Jacob Landman had already made a major impact on the conversation about sterilization with the 1932 publication of *Human Sterilization*, in which he was critical of the scientific rationale for widespread compulsory sterilization.[29] His work was particularly well received by John Ryan, who praised Landman as having "done the work in a truly scientific spirit without exaggerating or minimizing or defending any particular thesis."[30] Given his reputation for fairness, Landman's contribution to the *Scientific American* debate represented the voice of moderation in the conversation; he provided a summary of what he viewed as the scientific shortcomings of the eugenic sterilization movement and an estimation of the contemporary status of that movement in the United States.

Landman began with the obvious: in 1934, the "questions of population restriction and selection have been revived because of the chronic conditions of unemployment." Depression conditions were compounded by the claim that nearly one-fifth of the population was supposedly "socially unadjusted or maladjusted" and consisted of the mentally diseased, the mentally deficient, the dependents, the delinquents, the degenerates, and the infectious. But Landman nevertheless cautioned against exaggerating scientific findings about this troubling fifth, arguing that only mental deficiency and mental disease had the possibility of being hereditarily transmitted. Since much of the justification for sterilization programs was purely conjecture, Landman called for better institutional care and considerably more scientific research on mental traits and heredity.[31]

In contrast to Jacob Landman's cautionary approach to heredity, eugenics, and sterilization, Ezra Gosney gushed with enthusiasm for the reformist possibilities that accompanied a far-reaching sterilization plan. As the president of the Human Betterment Foundation, the leading sterilization advocacy organization in California, Gosney wholeheartedly endorsed sterilization as a central element of social policy. He pointed to a problematic decrease in the birthrate among the "intelligent part of the population," and an increase in the birthrate among "the unfit." Gosney's prescribed remedy for this problem and its associated social ills was a program of aggressive eugenic sterilization. Claiming that the procedure had almost no side-effects for the patient, and myriad benefits, he championed the successes of the twenty-five-year-old California sterilization program, which he and Paul Popenoe had studied extensively through the Human Betterment Foundation. Though in many cases California patients and their families had voluntarily consented to the operation, Gosney discussed four situations that might necessitate state authorization. In the first instance, a patient might not have family members to represent him or her; in the second, existing relatives might have the same deficiencies as the patient; third, a patient's relatives might desire the operation but might not want to share their desire with the patient; and fourth, a patient's family might take "a view of the matter which can hardly be shared by their fellow citizens." In the latter case, said Gosney, the state would have the duty and the right to forcibly sterilize the patient even over the wishes of the family. Gosney endorsed this comprehensive eugenic sterilization program as "one of the many measures that are indispensable in any far-sighted and humanitarian program for dealing with society's tremendous burden of mental disease, deficiency, and dependency."[32]

One month after Gosney's piece appeared in *Scientific American*, the magazine published an article, authored by Dr. C. Thomalla, which they had obtained through the German embassy in Washington, DC. Because of its source, the editors urged their readers to view the article as "an officially sanctioned statement of Germany's aims in connection with racial and national eugenics." Thomalla's piece immediately affirmed the connection between the eugenics movements in the United States and in Germany. Discussing the alarm provoked by the German eugenic statutes and the charges leveled by human rights advocates around the world that the German law represented "a return to barbarity and to the unchristian customs of the Huns," Thomalla noted that his country accepted that charge "jointly with those 27 enlightened and progressive states of America." Thomalla then turned to justifications for eugenic sterilization regulations, the primary one being economics. "In a state which is desperately fighting for its existence, as is Germany," Thomalla explained, "it is a simple problem of arithmetic whether to allow for all eternity the procreation of defectives, thereby most heavily burdening financially the working people, or whether to interfere before it is too late with the career of the worst and most costly of this kind of defective." In addition to the economic demands of "this kind of defective," Thomalla echoed Gosney's point about differential birthrates as a reason for increased urgency in eugenics policies. In light of such urgency, Thomalla dismissed religious objections to eugenic sterilization by suggesting that "hyper-civilization" allowed for the increase in defectives that harmed the rest of the population. "National Socialist Germany," he claimed, "considers interference with unfit life to be a sound application of the true Christian love of one's fellow man."[33] This willingness to ruthlessly sacrifice individuals in the service of the common good justified not only forced sterilizations but so many of the other horrors perpetrated by the Third Reich.

In response to these glowing endorsements of eugenics sterilization as sound social policy, Ignatius Cox offered a multifaceted condemnation based on scientific research and natural law reasoning. In his article, entitled "The Folly of Human Sterilization," Cox laid out the findings of a small but growing number of American biologists who rejected the hereditary inheritance argument that was foundational for much of the eugenics movement. Cox cited the work of the biologist Raymond Pearl, the psychiatrist Abraham Myerson, and that of Jacob Landman to argue that environmental factors often far outweighed heredity in causing and treating such deficiencies. Even more important was the work of the geneticist Herbert Spencer Jennings on gene theory and the inheritance of human characteristics. Jennings cautioned biologists that family pedigree and intelligence tests in lieu of rigorous experimental research did not provide a satisfactory explanation for the transmission of human characteristics. A sterilization program based on simple gene theory could not hope to effectively eliminate all of the carriers of recessive genes for feeblemindedness, if feeblemindedness even followed a simple gene pattern. Taking Jennings's cautions to heart, Cox extended the argument by speculating that the unknown rate and cause of genetic mutation might make sterilization efforts futile. In the case of eugenics, Cox explained, "emotion has been substituted for reason, and wishful thinking for scientific procedure." To this scientific perspective, Cox added a concise summary of the Catholic Church's natural law arguments against sterilization. Affirming his scientific and moral arguments, Cox closed by suggesting that if eugenics advocates truly wanted to improve the human condition, they should work to eliminate war and social injustice. By developing his argument in this way, Cox moved beyond the approach taken by Catholics in the 1920s who had hewed closely to scientific and legal arguments in their public opposition to sterilization laws. He was confident enough of his place in the American community that he felt comfortable framing the Catholic teaching for a popular, non-Catholic audience.[34]

In emphasizing the work of Raymond Pearl (1879–1940) and Herbert S. Jennings (1868–1947), Cox tapped into a group of US and British scientists, none of whom were Catholic, who vigorously disputed the scientific underpinnings of what the historian Daniel Kevles has termed "mainline" eugenics. In contrast to Charles Davenport, Harry Laughlin, S. J. Holmes, Paul Popenoe, Ezra S. Gosney, and the other champions of eugenic social ideology, these men disputed the racist and classist implications at the heart of the eugenics agenda, even as they continued to support some version of eugenics science. Jennings had come to public attention with his outspoken criticism of the xenophobic impulses behind the 1924 immigration restrictions. Having studied with Charles Davenport at Harvard before joining the faculty at Johns Hopkins University, Jennings was until the early 1920s firmly associated with mainline eugenicists. His reaction to the racialist assumptions of the National Origins Act distanced him from that core group. Then, in 1930, Jennings produced a successful and popular book, *The Biological Basis of Human Nature*, which undercut the generalizations he saw as the backbone of eugenics science.

Also at Johns Hopkins, Raymond Pearl, a professor of biometry and vital statistics, was one of the first eugenicists to openly dispute some of the movement's assumptions about the role of heredity and environment. In "The Biology of Superiority," published in the November 1927 issue of the *American Mercury*, Pearl rejected the reactionary tone of eugenics advocates like Madison Grant, Lothrop Stoddard, and Harry Laughlin, whose racial and class prejudices polluted objective science. Though these early conflicts with the core scientists of the eugenics movement set Pearl and Jennings apart as reformers, there is some question as to the depth of their antiracist sympathies. In comparison with their British reformer counterparts—especially J. B. S. Haldane, Julian Huxley, Lancelot Hogben, and Lionel Penrose—Pearl's and Jennings's critiques of racism seemed mild. Each in his own way continued to function with some baseline acceptance of scientific racism: Pearl would often privately voice a vicious personal anti-Semitism, and in his work Jennings cited approvingly Charles Davenport's book *Race Crossing in Jamaica*, which condemned miscegenation. Despite these contradictory perspectives on eugenics and racism, their critiques of the eugenics movement created an opening for other critics, such as Catholic intellectuals and social scientists, to oppose eugenics policies.[35]

In the spring of 1935, given the heightened public interest in issues surrounding sterilization and the number of scientists and physicians who were stepping forward to question the efficacy of eugenic sterilization, the editors of the *Forum* decided to "organize a popular, non-technical debate on the subject." Henry Goddard Leach approached John Ryan and Paul Popenoe to present opposing viewpoints on sterilization. In making his request of Ryan, Leach forwarded a manuscript of Popenoe's piece, urging Ryan to represent the "opposition party." Paul Popenoe set the tone for the debate by describing sterilization as a relatively minor surgical procedure with only positive implications for the patient, his or her family, and society. He told his readers that the ability of sterilization to prevent the birth of significant numbers of defective children, or of those who would be raised in "undesirable" surroundings, was "enough to justify its inclusion as one of the measures indispensable to any intelligent and effective program of social welfare."[36] Ryan responded with an essay entitled "Futile Immorality," which in fact presented some of the technical aspects of the question that had recently been raised by critical eugenicists.[37] In this respect, Ryan acknowledged the scientific veneer of the campaign for eugenic sterilization, the critique provided by particular elements of the scientific community, and the importance of presenting both to the general American public.

While the debates in *Scientific American* and the *Forum* brought the Catholic opposition to eugenics to a secular audience, the leadership and staff at the NCWC continued to insist that clergy and laypersons alike have adequate and up-to-date resources in presenting both the moral and scientific material that formed the basis of the Catholic opposition to sterilization. In August 1934, Frank Hall, the director of the NCWC News Service, wrote to John Burke about producing yet another series of articles dealing with sterilization, which he hoped could then be compiled in a pamphlet. Hall secured Dr. James J. Walsh to write the pieces dealing with the medical and scientific issues, but he was still looking for someone to deal with the moral issues. When he approached John Ryan, he discovered that the head of the Social Action Department was far too busy. Therefore, Hall inquired as to whether Burke thought Ignatius Cox would be a suitable person to work on the project. Hall had a copy of Cox's article from *Scientific American*, and he speculated that Cox "would be inclined to accept a commission to write the articles in question, and also that his inclination would be to deal with the points we have in mind." Burke agreed, and the resultant pamphlet, entitled *What About Sterilization?*, contained eight essays, five from Walsh and three from Cox.[38] These eight small essays joined a body of material from the NCWC dealing with marriage, family life, birth control, and sterilization. Of the available resources listed on the back page of the pamphlet, thirteen of the twenty-four items dealt substantially with sterilization or eugenics.

Thus, by the middle of the 1930s, the Catholic press, clergy, and the NCWC had succeeded in bringing their position not only to a Catholic audience, but to a non-Catholic one as well. The presence of Catholic authors in the features produced by *Scientific American* and the *Forum* allowed a secular readership to become familiar with the full range of Catholic arguments in opposition to eugenics. They presented both scientific and natural law arguments. In doing so, they balanced moral concerns with scientific skepticism. The NCWC News Service continued to disseminate the most developed, articulate, and scientifically up-to-date information to American Catholics. In all of these venues, Catholic writers spoke with a confidence that was supported by the increasing scientific evidence against the effectiveness of negative eugenic measures and the growing public recognition of Catholics as the primary voice of opposition to such policies.

A Vocal Few

The success in voting down forced sterilization legislation in Ohio was not a complete surprise given the large number of Catholics and their influence in that state. But Catholic lobbying efforts were vigorous as well in states where they were in the minority. In the 1930s in the Deep South, for example, sterilization bills came before the legislatures of South Carolina (passed in 1935), Georgia (passed in 1937), and Alabama.[39] Of these three states, only Alabama had a concentrated and active Catholic community. With close to 33,000 church members, Catholics in Alabama represented only a small fraction of the total population of roughly 2.7 million people. However, given the extensive informational material dealing with eugenics and sterilization available from Catholic sources coupled with the steady stream of coverage in the Catholic and the secular press, this relatively small Catholic population base was able to vigorously oppose the sterilization bill introduced by Representative Aubrey Dominick and Senator Hayse Tucker in 1935.[40]

The bill would have covered a broad swath of society, authorizing the forced sterilization of the insane, the feebleminded, and habitual criminals as a condition of release from prisons or institutions, and the sterilization of the mentally ill and mentally retarded on the recommendation of the county medical society.[41] Supporters argued that the hereditarily deficient were "vicious," "useless," and ultimately unproductive members of society. Religious opposition proved to be key in the debate, with Alabama's Baptists and Methodists weighing in on the bill in addition to Catholics. Additionally, the American Civil Liberties Union and local trade unions also objected to the legislation. In the end, however, procedural issues proved as important to the legislation's fate. Despite being overwhelmingly in the minority, activist Catholics concentrated in Mobile and Birmingham aggressively lobbied their local legislators and helped to influence Governor Bibb Graves to veto the bill. In waging their opposition, Catholics drew on the latest scientific material available, but they also conjured up the specter of an encroaching state, cited the overriding influence of the Depression on fiscal concerns for the bill's popularity, and noted an unsettling commonality between the Alabama bill and Germany's sterilization policies.[42]

Introduced in the spring of 1935, the Dominick sterilization bill was the subject of vigorous debate on the floor of the legislature into the late summer. The bill had the backing of the state health officer, institutional superintendents, and the local press. Dr. William Partlow, a physician in charge of the Alabama state hospital system for the insane, was the bill's leading institutional supporter. As early as February 1935, the *Montgomery Advertiser* had expressed support for the bill, reproducing a statement from Dr. J. N. Baker, the state health officer, and arguing that the "vicious and the useless should not be allowed to reproduce their kind," and that the procedure would not actually inflict injury on those subjected to the measure. It seemed, early on, that the Dominick bill had a good chance of making its way through the legislative system to the governor's desk.[43]

As the Alabama bill seemed to gain support from mental health officials and newspapers around the state, Catholics began to take notice. Speaking at a social workers forum in April 1935, the Jesuit priest William F. Obering derided the tendency of social workers to approve of sterilization laws. A professor of ethics and sociology at Spring Hill College in Mobile, Obering's views were based on standard elements common to Catholic opposition: the infringement of the state on human rights, the presence of social dangers (presumably a rise in nonmarital sex facilitated by absence of the threat of pregnancy), and the ineffectiveness of sterilization. William Partlow offered a counterpoint statement to the forum, in which he argued in favor of sterilization as a tool for effective social work.[44]

In early May, Obering published a long article in Birmingham's *Catholic Week* expressing his views on the Dominick bill, which was to come before

Alabama's full house of representatives that month. He argued that the philosophy underlying the sterilization statute violated the rights of citizens and rendered them "mere chattel of the State, not differing essentially, as far as [their] juridical position is concerned, from the animals of the State experimental farm." While one might suppose that Obering's mention of "chattle" was a reference to the history of slavery and racial discrimination in the South, he actually constructed his discussion around the classical case of Roman slavery. He pitted the model of Roman slavery, in which the slave had no rights, against the vision of inalienable rights voiced in the Declaration of Independence. Obering pointed to the philosophies of Hegel and Marx as modern versions of the Roman principle in which the citizen was subjected to the state and in which God was displaced. He submitted that such an elevation of the state or the people over the rights of the individual was the driving force behind "Hitlerism in Germany, with its blood purges, its shooting fests, and its deliberately planned famine." By making such a dramatic comparison, Obering set the stage for a protracted and heated legislative battle over the measure. At the same time, in his tone and in his invocation of the unjust imposition of state power, Obering showed that he understood the broad grounds on which Catholics, although a minority within the population, could oppose eugenic sterilization.[45]

Despite Obering's efforts, the Catholic opposition was, quite simply, too concentrated in a few urban areas to influence sufficient numbers of local representatives across the state. The Dominick bill was passed by Alabama's house of representatives by an overwhelming margin. But, while lopsided, the vote was not without protest. The representatives from Mobile, the area of highest Catholic concentration in Alabama, led the fight against the measure through a procedural effort to postpone consideration of the bill indefinitely. Though Mobile Representatives Taylor and McDermott were unsuccessful in their attempt to delay the bill, the Alabama Knights of Columbus registered their complaints about the proposed legislation by submitting a resolution into the record. In contrast to this Catholic opposition, the statute received the support of the medical community and local newspapers.[46]

With widespread public support, the bill moved on to the state senate for consideration in late May. In hopes of influencing the reception of the sterilization bill in the senate, a Catholic delegation testified before the Public Health Committee on May 15, 1935. The twelve-person group, composed of doctors, scientists, lawyers, priests, and other interested laymen, presented a multifaceted critique of the bill, despite the fact that only three of the men got to testify. Together, E. T. O'Connell, the editor and publisher of Birmingham's *Catholic Week*, Dr. A. L. Stabler, the State Deputy of the Knights of Columbus, and William Obering registered a number of complaints with the committee, and they reminded the senators that the opposition to the measure came not only from Catholics but from non-Catholics as well. Cautious about the fiscal demands of the Depression, they also speculated that the law was being advanced on economic grounds rather than on purely scientific grounds because it would allow for release of large numbers of institutionalized persons. Thus, the delegation tried to point to the small but growing coalition of critics who opposed eugenic sterilization. In making their arguments, the delegation cited liberally from relevant court cases and new scientific opinions on sterilization.[47]

Despite the multifaceted range of objections raised by the Catholic delegation, the senate's Public Health Committee sent the sterilization bill to the floor for a vote on May 31, 1935. On the main floor, Senator Tucker shepherded the law through a lengthy and caustic debate that lasted close to three hours. Senator Rogers, of Mobile, led the opposition to the statute, speaking for more than an hour on the matter. In the course of the debate, the senators focused their discussion on two key issues: the relationship of religion to science and the procedural provisions of the law.

Though many senators commented on the widespread support of the medical and scientific community for sterilization, others framed the debate in terms of religious questions, citing biblical principles and issues of divine creation. Though Catholics were a small minority in the state, Baptists and Methodists were abundant. In fact, there were six times as many Baptists and four times as many Methodist as Catholics, and some among them had reservations about the bill.[48] For example, Senator Woodall read into the record letters of opposition from a leader of the Alabama Baptist community and from the *Alabama Christian Advocate*. Referring to the well-known Catholic opposition, Senator Browder declared that the church "has always stood in the way of scientific progress." Senator Rogers then countered Browder by pointing to Mendel and Pasteur as notable Catholic scientists. More substantially, however, Rogers, in what the *Montgomery Advertiser* called the "most determined and scholarly" objection to the law, argued that the bill was unconstitutional because it provided no venue for appeal. On this matter, Rogers was supported by the ACLU and local trade unions, who worried that the bill was so far-reaching that it was ripe for abuse. Nevertheless, at the close of the debate, the measure passed, 18 votes in favor to 9 against.[49]

Much to the surprise of Alabama political insiders and supporters of the sterilization bill, Governor Bibb Graves vetoed the legislation in late June. Graves based his veto on the advisory opinion of the state's supreme court, which counseled that the bill was technically unconstitutional because it failed to provide for due process of law. Graves also mentioned the divided opinion of scientists: "I am convinced that until the scientists themselves reach further agreement and until we are in a position to supply to all concerned such legal and medical and county expenses as would be necessary for the protection of their individual rights, this is not a wise policy for Alabama to adopt." The *Mobile Register* noted that the decision had come one day after a delegation from Mobile, including a number of Catholics, had appeared at a public hearing.[50]

The governor's veto satisfied Catholic lobbyists, and also found favor with influential Baptists, such as the prominent minister Lesley Gwaltney. While Gwaltney was not a theological liberal, he was a major supporter of social Christianity. Writing after the veto, he informed *Alabama Baptist* readers that "no mortal on earth, sane or insane, white or black, ought to be forced against his or her will to [be sterilized] without any means of redress."[51] In doing so, Gwaltney gave support to Senator Roger's concerns about due process and the constitutionality of the statute.

Eugenics supporters, however, were not deterred. Rather than challenging the veto, Representative Dominick immediately introduced a new bill that he hoped would satisfy the procedural concerns of the supreme court. The new bill progressed through committee and both houses of the legislature in much the same manner as the original, only to face another veto from Graves. In a more vigorous response, Graves explained that the sterilization process entailed an unacceptable risk of death for female patients: "Those who will die are innocent and pure, have committed no offense against God or man, save that in the opinion of experts they should never have been born. . . . Simply because it may be cheaper to operate on all and kill some than it is to segregate all, still I insist that we haven't the right to kill a single one of these innocents."[52] Though William Partlow speculated that Catholic pressure had been the primary influence in killing the bill, there were a number of additional factors at work. Graves's own religious views as a member of the conservative Church of Christ may have played a part. And there was also the opposition of organized labor. Moreover, Graves had been elected in 1926 with wide support from the Ku Klux Klan, but during the Depression years he had embraced New Deal reform. Thus, the governor's personal and political history left him with a complex set of commitments, all of which could have played a role in his decision to veto the bill. Representative Dominick considered trying to override the second veto, but he was unable to gather enough support and decided to accept the governor's decision.[53]

Though sterilization supporters in Alabama tried to pass similar measures in 1939 and 1945, they never regained the momentum of 1935. The tide of support for sterilization turned from enthusiasm to caution. This was due in no small part to the diligent and outspoken opposition of Alabama's small Catholic population. By following the lead of Catholics in other states in voicing opposition on scientific, legal, and humanitarian grounds, the Catholic delegation from Mobile was able to find individuals with similar objections in the Baptist and Methodist communities as well as in labor organizations and the ACLU. The confidence with which they communicated their opposition contributed to the growing recognition by the general public that Catholics were the central voice in a growing chorus of scientists, humanitarians, and social activists who opposed sterilization.

The Face of Public Opposition

All of this activity, and the confidence of its tone, insured that throughout the 1930s Americans of all stripes were more and more likely to see Catholics as a useful source of information about sterilization. John Ryan's correspondence in the 1930s, for example, reveals that non-Catholics consulted him for guidance on the church's teachings about sterilization and eugenics.

In 1934, Mary Gawthorpe, the prominent British suffragist who had emigrated to America in 1916, approached Ryan as an ally in the fight to block sterilization statutes. Gawthorpe considered herself to be politically aligned with Ryan due to their common interest in labor and minimum-wage issues, so she felt comfortable contacting him with her concerns about the economic motivations of sterilization advocates. She questioned Ryan as to whether the time was not ripe "for a special study from the angle of industrial ethics to find out as dispassionately and in as unbiased a way as possible (for if there are to be any 'breaks' I claim that the derelicts, the economic 'failures,' should have them) what is really at back of the so called 'eugenics' of today and where is this most diabolical kind of prohibition leading?" With this query, she enclosed copies of three letters to the editor that she had written to New York newspapers regarding eugenics and sterilization. Gawthorpe's interest in industrial relations aligned her with Ryan in trying to solve economic problems rather than resorting to eugenics policy initiatives that did not fundamentally change the causes of poverty. Ryan heartily agreed with her concerns and sent back a stack of materials to aid in her opposition efforts.[54]

As Catholics achieved more recognition for their role in the eugenics debate, they expanded their vision for how the government should treat its citizens. Long advocates of charity relief, the dire circumstances created by the Depression only amplified the Catholic presence in welfare work. But while Catholics took care of their own, just as they had always done, they now called on state governments to follow their lead. Elected officials, they insisted throughout the 1930s, should give more help to their constituents, but at the same time, those officials needed to be vigilant in protecting the rights of citizens to be free from unnecessary government intrusion, bodily and otherwise. The ongoing willingness to pursue this balance is further evidence of Catholics' increasingly confident voice in the political arena and in the pages of the secular press.

The prominence of the Catholic opposition to sterilization measures after 1930 did not escape the attention of eugenics supporters. The March 1946 issue of *Eugenical News* contained a survey of the status of sterilization legislation around the country. By that date, there were twenty-six active sterilization statutes and a total of six state laws that had been found unconstitutional. Authored by Marion S. Olden, then the executive secretary of Birthright, Inc. (formerly the Sterilization League of New Jersey), the article discussed the main roadblocks to a nationwide program for the sterilization of hereditary defectives. Olden told her readers that in the campaign to promote legalized sterilization, "the opposition of Roman Catholic leaders constitutes the greatest obstacle that is encountered in applying, or in acquiring this therapeutic protection." Echoing the findings of her research from 1935, she followed this comment with a list of instances in which eugenics advocates cited Catholic efforts to thwart the passage or implementation of sterilization laws. In all, she noted there had been effective Catholic opposition, in the form of vocal political pressure from both clergy and the laity, in Wisconsin, Alabama, Colorado, Wyoming, and Pennsylvania. At the same time, she reported that in Maine, Arizona, Connecticut, New Hampshire, and in some areas of Michigan, Catholics had provided enough pressure that the use of existing and legal sterilization provisions was being curtailed. Olden and her fellow eugenics advocates had faced Catholics as political adversaries for so long that she did not even bother to explain the reasons for that vociferous Catholic opposition.[55]

By making themselves more visible throughout the 1930s, in the legislative debates about sterilization and in the pages of the American secular press, Catholic intellectuals and activists concretized the association between Catholicism and sterilization opposition in the minds of non-Catholics, Catholic laypersons, and eugenics advocates. The attention to the sterilization movement and the scientific critiques of it, in combination with the participation of Catholic intellectuals in secular debates, contributed to a more widespread opposition to eugenic sterilization. The work of Catholic intellectuals in both the Catholic and non-Catholic press added to the resources available to laypersons who wanted to actively oppose legislation in their state legislatures. In turn, the widespread public exposure of the Catholic position attracted the attention of non-Catholics who had questions about the issues surrounding the movement. Finally, eugenicists came to grudgingly accept that US Catholics had proved to be vocal and dedicated opponents whom they could no longer ignore.

SIX

A Great, Popular, Noncontroversial, and Effective Movement: Struggling with the "New Eugenics"

In 1939, Frederick Osborn, president of the American Eugenics Society, stood to address the delegates to the Conference on Relation of Eugenics and the Church. He informed them that on that subject he was "an optimist by nature." However, after he had heard the four papers from the speakers on the panel that day, he was more than a little disappointed. In fact, his dismay was so great that he proclaimed himself "more bitterly discouraged than I have ever been in the years I have been working and thinking about eugenics." His dismay was due to the fact that the speakers had failed, in his estimation, to account for the changes in the eugenics movement over the previous twenty years.[1]

Contrary to the clergymen on the panel, who had critiqued the immoral vision of the eugenics philosophy, Osborn believed the organization of which he was president was promulgating a "new eugenics." The AES, he was proud to say, was adjusting to the scientific advances that questioned the simplistic understanding of human heredity advanced in the early decades of the eugenics movement. The organization was moving toward a program that relied less upon sterilization as a silver bullet for degeneracy, and more upon positive eugenics. Given this shift in emphasis, Osborn hoped that a meeting between eugenicists and the clergy might make evident their common mission to "teach people the values of family living."[2] He was convinced that the members of the clergy would share his concern that "births have declined to a point far below the number required for replacement among the more responsible church-going people in all our cities."[3] These ideas, however, seemed peripheral to the interests of the speakers.

Convened in May of 1939 at the Town Hall Club in New York City, the conference was attended by quite a number of prominent Catholics, both from the New York area and from Washington, DC. Among the conferees was the counselor to Catholic students at Columbia University and a representative from Catholic Charities. The Reverend John Montgomery Cooper, the Reverend John O'Grady, and Monsignor John Ryan, from the Catholic University of America, were also present. Finally, in attendance was the Reverend Edgar Schmiedeler, representing the NCWC's Family Life Bureau. Before this distinguished crowd, the Reverend Francis J. Connell (1888–1967), a moral theologian from Mount St. Alphonsus in Esopus, New York, delivered a paper entitled "The Catholic Church and Eugenics" that succinctly expressed the perspective most frequently argued by Catholics in opposition to the heart of eugenic thinking.[4]

In his address, Connell focused on the three basic principles of Catholic teaching that structured the standard reasoning about eugenics policies: first, that "the soul of man is immeasurably more important than his body"; second, that "the primary end of marriage pertains to the common good of society, for it is the procreation and the bringing up of children for the maintenance and propagation of the human race"; and third, that "there is an objective code of morality established for the regulation of human conduct by the Almighty," so that "even when one's ultimate aim may be to effect some genuine good, the laws of God may not be violated."[5] Based on these three principles, Connell argued against birth control, sterilization, and public authority over marriage. His second principle pointed to the space where Catholics and eugenicists had traditionally found common ground: encouraging strong, baby-filled nuclear families. Nonetheless, a tension remained over the selectivity of eugenic pronatalism and its tendency to advocate birth control as a way to reduce "unfit" reproduction.

Despite Connell's familiar emphasis on birth control, sterilization, and marriage restriction as the core of the eugenics movement, Frederick Osborn did have a point: starting around 1935 and continuing into the 1940s, the eugenics movement in the United States had dramatically changed its focus as it adjusted to the new scientific and cultural thinking about heredity and environment. Eventually, though not quickly enough for Osborn, Catholic conversations about eugenics and eugenics policy initiatives shifted in response to these changes. As eugenics organizations backed away from promoting involuntary sterilization, eugenicists began to promote positive eugenics through marriage counseling that emphasized traditional understandings of family life. During this period, Catholic activists also focused on traditional family life and reproduction. The hardships of the Depression and the war years produced concern among both Catholic activists and eugenics supporters about the health and survival of American families. Though they approached these issues with different priorities and concerns, Catholic family advocates and eugenicists found some accord in their pronatalism, particularly by the mid-1940s with the emergence of the "baby boom." Despite this apparent agreement about the "family crisis," Catholics and eugenicists continued to find themselves on opposite sides of important questions about the role of the state in regulating and restricting marriage. Thus, the era just before and after World War II witnessed a surprising combination of clear clashes and subtle similarities over who should be encouraged to marry and have children.

The New Eugenics

During the mid-1930s, given the growing cultural and scientific pushback against involuntary eugenic sterilization, the eugenics movement sought to redefine its public face by backing off from its earlier position on forced sterilization. Convinced that the AES needed to distance itself from its more controversial public policy initiatives, the leadership sought to refocus its agenda. The result was a repackaging of eugenics philosophy to embrace environmental reforms, encourage high birthrates, eschew past racism (at least in official rhetoric), and to become less strident in its call for compulsory sterilization.

Frederick Osborn (1903–1980) led the AES through this reformation during the 1930s. A Princeton-educated businessman, Osborn retired in 1928 at the age of forty to concentrate on the study of eugenics. His uncle, Henry Fairfield Osborn, secured him office space at the American Museum of Natural History and shortly thereafter he began his involvement with the AES, the Eugenics Research Association, and the Galton Society. While Osborn supported the widespread use of sterilization and admired Laughlin, Popenoe, and Gosney, his primary focus was on "practical eugenics." In the application of eugenic principles, he encouraged a view of intelligence that was determined on an individual basis through IQ testing rather than on the basis of race or class groupings. This stance allowed Osborn to distance himself from the more overt racism of earlier eugenicists, but it failed to recognize the biases that were built into the IQ examinations, which continued to perpetuate notions of racial differences in intelligence. Nevertheless, his focus on individuals allowed him to present a new vision of eugenics that seemed more humane and more nuanced than previously.

By 1935, Osborn and his allies had control of the AES and he set about reworking the organization by revising its constitution and streamlining its internal structure. The result was a slight increase in membership and financial stability.[6] Osborn's leadership also signaled a move toward a more positive agenda that recognized the importance of the environment for those with normal heredity. His statement, "A Eugenics Program for the United States," drafted and released in 1935, publicized this shift.[7] Though Osborn's work marked a new direction for the AES and attracted some new adherents, the organization still failed to attract the kind of widespread public support that it had enjoyed in the late 1920s. In May 1938, Osborn addressed AES members at the organization's annual meeting, discussing what he felt was an anemic public response to the changes in the organization. He speculated on why the public had developed "false impressions" of the eugenics movement, arguing that the AES needed to take those impressions into account in planning for the future of the movement. He pointed out that class and racial bias had motivated the eugenics movement in the past, both at home and around the world. Osborn cautioned against giving any impression that the contemporary movement was founded on such biases. He then suggested that the movement had been confused with the sex hygiene, sterilization, and birth control movements, all of which were to some degree associated with eugenics but which did not represent the sum total of the eugenics program. Under his watch, Osborn asserted, the AES had instead "been trying for three years to associate eugenics in the public mind with housing, with recreation, with public health nursing, and with all phases of education."[8]

Despite these efforts at reorientation, the AES had met with little popular success. As a consequence, Osborn proposed a further narrowing of the organization's message to differentiate it from all other organizations and movements. The AES needed to focus on raising the birthrate of "above-the-average people" and on encouraging the use of birth control by the unfit. Osborn saw this plan as "the most direct way in which eugenics can become a great popular noncontroversial and effective movement, with wide public support."[9] This continued distinction between the fit and the unfit did not mesh with the near universalism of Catholic pronatalism, but it did point to some limited space for agreement about encouraging a more vigorous birthrate. Though Osborn's statements about promoting eugenics as a "popular noncontroversial" movement suggest that he wished to rescue the AES from its previous association with controversial issues, he still was not suggesting that the organization actually abandon its stance on negative eugenics.

Rather, Osborn went on to suggest that the eugenics movement make a move similar to the shift that was underway in the birth control movement: they should place the controversial elements of the discussion in the hands of medical professionals, not private organizations. He argued that the AES should "make every possible quiet effort" to encourage scientists and physicians to take up their responsibilities to the movement. In addition to his concern about research on heredity, he had a more concrete goal: "The public will accept negative eugenics from the doctor in a way it would certainly not accept it from an organized but nonscientific movement. I believe broad and wide use of sterilization can get acceptance through the medical profession quietly and without controversy."[10] Thus, Osborn supported a campaign to publicize and promote the new "noncontroversial" agenda: the AES would become a clearinghouse for eugenics information, leaving the controversial work of encouraging eugenic sterilization to go on behind the closed doors of doctors' offices.

This shift in focus did not represent a major shift in the ideological orientation of the organization, but rather a reckoning with the growing public antipathy toward the negative aspects of the movement. By supporting widespread access to voluntary sterilization through private physicians, the AES and other eugenics organizations would not have to address concerns about the appropriate role of the state versus individual rights. Osborn's impressions about fading public support for involuntary eugenic sterilization were confirmed in subsequent years.

During the war years, for example, sterilizations performed in state institutions declined dramatically as institutional staffs were trimmed for the war effort. According to Philip Reilly's account of involuntary sterilization in the United States, between 1941 and the end of 1944, the number of sterilizations performed was only 40 percent of the number performed between 1935 and 1940.[11] At the same time, much of the old guard in the movement had passed away. In 1940, the Carnegie Institution had withdrawn funding for the Eugenics Record Office; and in 1942, with the death of Erza Gosney, the Human Betterment Foundation had closed.[12] Then, Harry Laughlin died in 1943, and Charles Davenport passed on in 1944. This left Osborn's AES and Birthright Inc. (Marion Olden's organization) as the major organized elements in the eugenics movement. Eventually, Birthright Inc. transformed itself into the Association for Voluntary Sterilization, and in the postwar years it promoted general access to sterilization as a means of contraception, especially for the "unfit."[13]

Opposition to eugenic sterilization was a key element of Catholic teaching on family life, but it was only one factor in a complex of issues concerning American family life in the first third of the twentieth century. Osborn's efforts to change the AES were undertaken in that larger context. Coinciding with the publication of *Casti Connubii*, the NCWC Social Action Department formed the Family Life Bureau in 1931 to address the special conditions that threatened the modern family. Placed under the control of the Reverend Edgar Schmiedeler (1892–1963), the FLB published *Catholic Family Monthly* and organized conferences to promote Catholic teachings on the family. A Benedictine priest trained as a sociologist at the Catholic University of America, Schmiedeler ran the FLB almost single-handedly, taking on speaking engagements and producing a torrent of print material to promote the agenda of the Bureau.[14] One bishop explained the centrality and importance of the Bureau's work for the rest of Catholic activism: "The most extensive and at the same time the most fertile field of Catholic Action in America and the world today is not the reconstruction of industry, or politics, or even in the schoolroom, but in renovating the Christian Spirit in domestic society."[15] Together, the National Council of Catholic Women, the National Council of Catholic Men, and the FLB sounded the warning on the status of family life in America.

This sense of alarm about the state of the American family was undeniably exacerbated by the Depression. For Catholics, the economic devastation of the early years of the Depression had created the deeply unsettling possibility of a restructuring of family life and gender roles in the United States. During the 1930s, the marriage rate in the United States dropped to an all-time low, birthrates declined sharply, and divorce increased. Desperate conditions also provided more opportunities for single women, as their participation in paid labor allowed them to become more self-reliant and more independent. As more and more men stood on the unemployment lines, their own masculinity, tied up with rigid gender roles, came into question. For some, the economic crisis allowed for a new sense of shared responsibility and hinted at a new equality of the sexes. But the historian Elaine Tyler May has argued that the Depression "also created nostalgia for a mythic past in which male bread winners provided a decent living, and homemakers were freed from outside employment." New Deal relief policies promoted this nostalgia by encouraging employment for husbands but discouraging it for their wives, and by maintaining wage differentials that reinforced women's dependence on men. In ways large and small, the unprecedented economic crisis placed a strain on marriages, and Catholics worried that the family would be forever altered.[16]

Two FLB pamphlets from the 1940s suggest that Schmiedeler was as concerned about the birthrate as Frederick Osborn and the AES were. First, in a text ominously entitled *The Threat of American Decline*, Schmiedeler sounded the alarm about the remarkably low population growth for the years between 1930 and 1940. While he discussed the role of immigration in population growth, the Benedictine's real target was the national birthrate. Though he recognized that the birthrate had been steadily declining for more than a hundred years, the dramatic decrease beginning in the 1920s was cause for concern. Schmiedeler warned of the effects of this population shift on schools, the economy, and social life. The population decline, he wrote, "shows a decadent nation, a selfish and undisciplined people. It shows the vicious, debilitating cancer of artificial birth control slowly draining away the vital powers of the people. It shows moral laxity, stamping in the spirit of pleasure and self-indulgence, leading to the shirking of hardship and self-restraint, producing a softness and deterioration that must prove suicidal to the individual and the nation. It shows the decay as well as the decline of America."[17] In Schmiedeler's eyes, such deterioration was almost guaranteed if Americans failed to embrace the Catholic perspective on marriage and the family, which called for a universal pronatalism. Significantly, Schmiedeler's formula for saving the family and the nation showed none of the selectivity of Frederick Osborn's plan, that only "above-the-average-people" would have more children.

In 1943, Schmiedeler wrote even more alarmingly about American culture and the family in *A Holy War against the Enemies of the Home and the Nation*. This publication ratcheted up his alarmist tone considerably. He used Catholic moral teaching to inveigh against a long list of social and economic conditions he believed inhospitable to strong families: birth control, abortion, divorce, infidelity, sterilization, improper parenting. Despite the array of forces turned out against Christian marriage, Schmiedeler explained that adherence to Catholic teaching could overcome all that threatened family life: "Dark as the picture may be, it would unquestionably be much darker had it not been for [the church's] efforts."[18] In this jeremiad, Schmiedeler took issue with those who would promote eugenics as a solution for the new economic strains that had been placed on the American family. But Schmeidler's text also echoed some of the alarm about the decline of the family that characterized eugenics texts. Thus it could be said that there was more common ground between his position and that of eugencists than ever before.

With the AES's turn toward the promotion of positive eugenics in the late 1930s, and with the focus on the family as promoted by the FLB, both organizations found themselves encouraging an even larger trend in American history that eventually came to be known as the "baby boom." During the 1940s, Americans re-embraced home and family life. In a marked break from the all-time low rates of marriage and births during the Depression, the war years saw an unprecedented increase in both marriage rates and birth rates. Between 1940 and 1945, the US birthrate increased from 19.4 births for every 1,000 adults, to 24.5 births. Similarly, the average age of marriage decreased, and the marriage rate increased, a trend that would remain consistent through the 1950s. The increasing tendency of Americans during the 1940s to get married earlier and have more children affirmed both the eugenicists' pronatalism and Catholic teaching on marriage and family. Elaine Tyler May maintains that this demographic anomaly cannot simply be explained by the peace and affluence of the postwar period; rather, she argues that a confluence of Cold War politics, therapeutic culture, and personal behavior rooted in a need for refuge and security produced the baby boom.[19] In addition to the psychological strains produced by Cold War politics and liberal consensus ideology, the role of therapeutic experts in this trend was key. The message of these marriage experts—psychologists, psychiatrists, family counselors, and social workers—was that couples needed to focus on the home as a source of strength and security in an environment that had been torn asunder by World War II and then by the strategic balancing act of the Cold War.[20]

To eugenics advocates, the new emphasis on domesticity was seen as the fulfillment of decades of work. The eugenics movement had emerging concurrently with the rise of the "expert" during the Progressive Era, and its supporters had long claimed that their scientific expertise in understanding heredity placed them in a unique position to forecast and manage the human population. Though the embrace of expertise in the 1940s centered on positive eugenics, it nonetheless affirmed the long-held stance of eugenicists that if they could convince the American population to heed their advice, they could revolutionize the race. Unsurprisingly, some of the marriage experts who came to the foreground during the baby boom, such as Morris Fishbein and Norman Himes, had an interest in eugenic philosophy.[21]

While Fishbein and Himes undoubtedly dabbled in eugenics, Paul Popenoe, a leading figure in the AES and an advocate of eugenic sterilization in California, emerged as one of the nation's most-well-known marriage advocates and counselors. A number of factors came together to make Popenoe the perfect man to promote eugenic marriage—in addition to the fact that he had married and fulfilled his own eugenic duty by fathering four sons. After World War I, Popenoe's eugenic activities included an attempt to enlist the help of middle-class men to ensure the eugenic improvement of the nation. In his first published work, *Modern Marriage: A*

Handbook (1925), Popenoe helped to popularize a language of scientific objectivity around the issues of marriage and family life for middle-class, mostly male, readers. In the book's introduction, he explained: "This book is . . . written from the biological point of view. It takes man as it finds him, and tries to make clear how he can fit himself into the American civilization of the twentieth century in such a way as to provide for his own greatest satisfaction and the progressive evolution of the race."[22] While Popenoe addressed his writing to the individual man, he made clear that each individual's choices had important consequences for the racial welfare of American society as a whole, drawing a link between race and nation and the public and private spheres.

Growing out of his efforts to encourage eugenic marriages in the mid-1920s, Popenoe founded the American Institute of Family Relations in 1930 to provide a center that would focus on education about marriage and the family. From the AIFR, which drew its initial funding from Ezra Gosney's Human Betterment Foundation, Popenoe became a pioneer in marriage counseling, eventually publishing numerous books and writing countless magazine columns on the difficulties of marriage and family life. By encouraging healthy middle-class Americans to forge strong marriages and large families, Popenoe believed he was reshaping the population based on eugenic principles. He accomplished this through a combination of education and counseling for couples with marital problems. By the early 1950s, the AIFR published a magazine called *Family Life*, employed thirty-seven counselors, and offered an advanced degree in marriage counseling. After World War II, Popenoe wrote less about eugenics, but essentially his views on positive eugenics remained the same.[23]

Not unlike his cordial relationship with John Montgomery Cooper, Popenoe expressed admiration for Schmiedeler's efforts at the FLB, particularly his efforts to encourage stable marriages that welcomed children. For example, after reading an article by Schmiedeler on parent education in 1934, Paul Popenoe wrote, "You people are really getting something accomplished, and all the rest of us envy and admire you." In return, Schmiedeler expressed his appreciation of the materials that Popenoe had forwarded from the American Institute of Family Relations. Schmiedeler included a copy of the *Catholic Family Monthly* with his letter. This mutual appreciation between Popenoe and Schmiedeler was centered on the efforts of both men to encourage the formation of strong nuclear families with plenty of children. Unfortunately, Schmiedeler's militant focus on supporting traditional families in the face of the upheavals of the Depression and World

War II led him to perceive an ally in a person who had spent his entire career championing negative eugenic measures that official Catholic teaching adamantly opposed.[24]

Regulating Marriage

Although Paul Popenoe and Edgar Schmiedeler agreed on their mission to promote strong, fertile, and lasting marriages, whether they recognized it or not, they remained far apart on other issues surrounding marriage. Popenoe had not only promoted eugenic sterilization during his career, but he also supported marriage restrictions. In addition to advocating sterilization as a negative eugenic measure, the AES was equally invested in the 1920s and during the Depression with preparing "model marriage" bills, which in turn attracted the support of anti-miscegenationists such as Madison Grant.[25] By the late 1930s, as the AES shifted away from a primary emphasis on negative eugenic measures, their investment in promoting "appropriate" marriage became even more important. Conservative Catholic moral theologians and more-radical Catholic activists, needless to say, resented the intrusion into what they viewed as the church's primary jurisdiction over sacramental marriage.

The state could rightfully legislate the civil effects of marriage, such as inheritance and taxes, but due to the sacred nature of the marital union, only the church held the authority to regulate and impose impediments to marriage. An example that clarifies this distinction is "social disease legislation." These eugenically inspired laws, common in the first half of the twentieth century, called for individuals to present a certificate stating that they were free from any sexually transmitted disease or other "defect" before the state would grant them a marriage license. If infected individuals could at a later date provide a certificate, signed by a physician, stating they were free from infection, only then would the state officially grant a marriage license. And therein lay the rub for Catholics: was this legislation creating a civil impediment to marriage, or was it merely creating a delay in the interest of public health?

Discussing a 1938 New York State regulation in the pages of *America*, Paul Blakely argued that the statute called for a justifiable delay in marriage that protected the public good rather than creating an impediment.[26] Others disagreed with Blakely's position on whether or not the state was overstepping its jurisdiction. Francis Connell refuted Blakely's position, based on the fact that the church claimed sole authority to regulate the marriage contract. Connell instructed readers that "since the Church has not legislated that social disease prevents a baptized person from contracting a valid and lawful marriage, Catholics must hold that there is no human legislation binding in conscience which directly prohibits a baptized person so afflicted from marrying." Additionally, Connell was wary that the legislation was simply a gateway, and urged his readers to be "alert to the danger which this present legislation may be only the opening wedge—the danger of legalized sterilization. The use of moral means, that is legislation, to prevent diseased persons from procreating may easily lead to the use of physical means."[27]

Connell's position on jurisdiction is instructive. He followed the letter of the law and tended to be suspicious of anything outside the elements delineated in that law, particularly in instances that involved a jurisdictional conflict between church and state. In Connell's opinion, social disease legislation and compulsory blood tests represented an attempt on the part of the state to legislate moral issues. "It is imperative therefore that Catholics be alive to the situation and realize that the civil government is now arrogating to itself a sacred power that Christ wished to be exercised solely by His Church, and that in passing eugenic legislation binding on the baptized the state is going beyond its lawful sphere just as truly as if it legislated as to who should be admitted and who should not be admitted to Holy Communion."[28] Connell was clearly convinced that the integrity of the church was at stake in these types of disputes. Confident though he was, the debate over social disease legislation continued into the 1940s in Catholic periodicals, drawing in questions and opinions from other notables, such as Thomas Vernor Moore.[29]

Social disease legislation was not the only eugenically inspired marriage restriction to draw the attention of Catholics. Antimiscegenation statutes eventually came under consideration, but only after decades of church concern about mixed religious marriages. Like other Americans caught in the historical and cultural legacy of white supremacy, many Catholics who worked for justice for African Americans within and outside the church, known as interracial activists, were not convinced that challenging such statutes would be beneficial. In writing about the issue of interracial marriage, John LaFarge, S.J., one of the founders of the Catholic Interracial Councils, showed a marked concern for the stress and hardship that contemporary American social conditions would place on those who entered into those unions. In *The Race Question and the Negro* (1943), he wrote that "racial intermarriage naturally produces a tension in family relations not unlike that tension which is produced by a mixed marriage in the field of religion" and so was "subject to a special impediment from the church."[30] By analogously linking mixed racial marriages with mixed religious marriages, LaFarge raised the issue of Catholic canon law.

This analogy between mixed racial marriages and mixed religious marriages made sense to Catholics. During the first three decades of the twentieth century, the "problem" of interfaith marriage had been a much more pressing issue than concerns about what Martha Hodes called marriage "across the color line."[31] The reasons were obvious. As a religious institution, the Roman Catholic Church traditionally sees the spiritual welfare of its adherents as its primary focus. Predictably, then, the Code of Canon Law explicitly regulates the conditions under which Catholics can enter into matrimony with persons outside of the faith, such as having an agreement to baptize and raise the children resulting from the out-of-faith unions as Catholics.

During the height of the second wave of immigration in the United States, in the years between 1880 and 1920, Catholics were concerned that new immigrants might leave the church as they assimilated into American culture—a problem commonly referred to as "leakage."[32] However, after the dramatic reduction in immigration in the 1920s, focus shifted to interfaith marriage as the cause of individuals leaving the church.[33] The language of racial difference sometimes haunted these meditations on interreligious marriage. For example, in a 1931 article filled with phrases like "color line," "mixed marriages," and "miscegenation," Joseph Donovan discussed the problem of the "invasion of Catholic life by the unregulated marriages of Catholics to non-Catholics." Entitled "Keeping Back the Color Line," the article was not a treatise on the changing racial dynamics of America's urban centers, but rather a meditation on what Donovan termed "spiritual miscegenation." The notion that interracial marriages were dangerous and negative was so pervasive in American life that he could invoke the specter of racial mixing to express his concern and distress about religious difference in marriage.[34]

In addition to the regulation on mixed-faith marriages, canon law as articulated in 1917 presented impediments based on age, impotence, an existing bond of marriage, sacred orders, and a number of other things. No such explicit statement, however, existed with regard to mixed racial marriages. Thus, there was never any barrier to interracial marriage in canon law. If Catholics were opposed to interracial marriages, they could not base that opposition on religious grounds. And they had no (religious) basis to support antimiscegenation statutes.[35] That is not to suggest that Catholics enthusiastically advocated interracial marriage, but rather that in a situation where such a marriage was a possibility, there would be no official cause for opposition. More than likely, most clergy would discourage such a union based on social conventions.

Statutes regulating intimate relations between whites and non-whites had been part of the fabric of the American legal system since colonial times. In the wake of the Civil War, African Americans had rushed to regularize their marriages in the eyes of the law since for decades slaves had been unable to contract marriage. Evidence suggests that both African Americans and whites were generally resistant to interracial marriages during this time, and since white men had limited access to black women, miscegenation decreased considerably when compared to the antebellum era. As Reconstruction gave way to disenfranchisement and Jim Crow, a complex of laws circumscribing African American lives—the antimiscegenation statutes—gained a prominent place in the law.[36] Progressively, the statutes began to use the trope of "blood quantum" to define a negro as any person who had any African ancestry at all. The so-called one drop rule worked to produce the appearance of an impenetrable wall of separation between whites and people of color. Among antimiscegenation statutes, the Virginia Racial Integrity Act of 1924 became the most famous. Passed in conjunction with Virginia's sterilization law, the Racial Integrity Act called for the State Registrar of Vital Statistics to ascertain the racial composition of every resident of the Commonwealth who wished to contract a marriage, and of any other person who desired to register. Based on that certificate of racial composition, it was then "unlawful for any white person in this State to marry any save a white person, or a person with no other admixture of blood than white and American Indian." The stated exception for American Indian blood allowed for a one-sixteenth or less admixture, in honor of the descendants of Pocahontas.[37]

If Catholic teaching was not officially opposed to interracial marriage, eugenicists pursued a broad-based agenda that often included overt support for antimiscegenation statutes. For example, Charles Davenport corresponded with W. A. Plecker, one of the chief proponents of the Virginia Racial Integrity Act, both before and after the act's passage in 1924, although he declined to offer assistance in the administration of the law. Additionally, on several occasions members of the AES leadership, such as Madison Grant, discussed various antimiscegenation statutes and their enforcement.[38] Similarly, Davenport crafted an extremely favorable review of Earnest Sevier Cox's book *White America*, which argued that the United States was in a state of racial deterioration that could only be halted by the mass migration of blacks back to Africa.[39] In 1929, Davenport continued in his exploration of interracial unions with the publication of *Race Crossing in Jamaica*, which he co-authored with Morris Steggerda. Studying the biological characteristics of the mixed-race peoples in Jamaica, Davenport attempted to show that they had below-average physical characteristics, in an effort to prove that heredity was stronger than environmental factors with respect to race. Davenport's imprecise classifications of races and groups, and his flawed interpretation of his data, garnered him a growing chorus of criticism on anthropometric and biological grounds.[40]

In the 1940s, as the understanding of race as a biological fact began to give way to the notion of race as a cultural construction, individual Catholics began to chafe at race-based marriage restrictions. For example, in 1945 *Commonweal* published a story by Iris Buaken, a non-Catholic married to a Filipino man, which detailed the resistance of Filipinos to California's antimiscegenation statute during the war and the support they received from the Catholic chaplain of the US Army's First Filipino Infantry. Unable to convince the California state legislature to alter its marriage laws, Chaplain Eugene Noury, the Red Cross worker assigned to the unit, and Colonel Robert H. Offley, made arrangements for Filipino servicemen and their brides to travel to New Mexico, where they were officially married, before they embarked on their mission in April 1944. Formalizing these marriages entitled the servicemen's families to sufficient allowances while they were away. A digest of the marriage laws in the Western states, along with an editorial note on canon law, followed Buaken's story. The editors maintained that antimiscegenation legislation was "in direct conflict with Canon Law." Hence, "a priest is in conscience obliged to fulfill his function as official and principal witness of such a marriage—in other words, he must 'perform' the marriage."[41]

Despite what the editors at *Commonweal* saw as a clear teaching, many Catholic interracialists remained unwilling to broach the issue of interracial marriage, fearing that any focus on this sensitive topic would cause a backlash directed at their work in other social and economic realms.[42] The reluctance to support interracial marriage was a testament to the power of racial taboos that involved sexuality and reproduction. Principled Catholic advocates for racial justice were hesitant to raise the topic, and those individuals who were less committed to interracialism were even more unlikely to call for equality in marriage. Nonetheless, just a few years later, the courage of a single interracial activist and a brave couple in California resulted in an affirmation of Catholic teaching from an unexpected quarter and made such resistance from individual Catholics far less tenable.

The Power of *Perez*

The potential implications of challenging antimiscegenation laws were vast. In his insightful work on the legal construction of race, *White by Law*, Ian Haney López emphasizes the role of law in structuring society, arguing that the law is "one of the most powerful mechanisms by which any society creates, defines and regulates itself."[43] Thus, López believes that in the construction of race, the law functions both as coercion and as ideology. It constrains action "through the promulgation and enforcement of rules that determine permissible behavior." But, just as important, it limits cognitive possibility by defining, "while seeming only to reflect, a host of social relations, from class to gender, from race to sexual identity."[44] By regulating the marriage contract and by advancing a definitive account of racial categorization, antimiscegenation laws epitomized both of these functions. In this way, the laws contributed to the effort to shore up white supremacy and the racial agenda of the eugenics movement. By 1948, thirty states had laws prohibiting the marriage of white persons to non-white persons, under various standards of definition. These regulations did not deal in the changeable status of sexually transmitted diseases, which were the focus of social disease restrictions, but rather in permanent, legally binding racial definitions.

As such, the laws drew the opposition of the National Association for the Advancement of Colored People and the American Civil Liberties Union. In 1913, under the leadership of W. E. B. Du Bois and in the wake of the rash of miscegenation statutes prompted by the black boxer Jack Johnson's controversial marriages to three white women and his arrest in 1912 for violating the Mann Act, the NAACP entered the fight over interracial marriage restriction. Stalwart opposition to new antimiscegenation statutes continued for decades and made the NAACP the most visible organization committed to marriage justice. In the 1930s, the ACLU joined the cause, echoing the NAACP's arguments about individual rights and marriage in their opposition to antimiscegenation statutes.[45] While these major civil rights organizations shaped the strategy for many civil rights battles to come, they were not major participants in the first successful overturning of a state law in the courts.

That role fell to the Los Angeles Catholic Interracial Council. In 1947, Daniel Marshall, a long-time activist at the council, agreed to help Andrea Perez and Sylvester Davis challenge California's antimiscegenation statute. The law prohibited state officials from issuing a license "authorizing the marriage of a white person with a Negro, mulatto, Mongolian or a member of the Malay race."[46] Marshall argued that since Perez and Davis were both members of the Roman Catholic Church, which maintained no official prohibition against interracial marriages, the California statute constituted a violation of their right of free exercise of religion by preventing them from participating in the sacrament of marriage. Marshall's argument in the *Perez* case was more than "an end-run strategy," as historian Peggy Pascoe has referred to it.[47] Rather, his appeal to the church's jurisdiction over the marriage contract was an attempt to bring an alternative, highly articulated, system of law into direct confrontation with a civil legal code that bolstered white supremacy.

Daniel Marshall saw the *Perez* case as a chance to directly pose the teaching of the church against America's system of legal racial oppression. One of the first steps Marshall took was to write to the auxiliary bishop of Los Angeles to seek support for the action. In an April 1947 letter, Marshall explained to Bishop Joseph T. McGucken his understanding of church teaching on the matter:

> When the State enacts laws inimical to the marriage laws of the Church, practically denying her right to protect the sacred character of marriage, she cannot allow her children to submit to such enactments. She respects the requirements of the State for the marriages of its citizen as long as they are in keeping with the dignity and Divine purpose of marriage. . . . The Church has condemned the proposition that "it is imperative at all costs to preserve and promote racial vigor and the purity of the blood; whatever is conducive to this end is by that very fact honorable and permissible."[48]

Marshall asked the bishop if he would meet with Perez and Sylvester to ascertain their readiness for marriage, and then whether he might testify in court in support of those elements of dogma that were essential to the case. Marshall must have been disappointed, if not surprised, by the response he received from McGucken. The bishop sent off a quick note of reply in which he chided Marshall for his presumptuousness, counseling: "I cannot think of any point in existing race relationships that will stir up more passion and prejudice than the issue you are raising. I doubt seriously the possibility of getting a balanced judgment in this matter.[49] In this response, McGucken's attitude reflected that of so many white Americans who might have felt some tinge of guilt about racial inequality but who were unwilling to push forward by causing turmoil in the service of justice.

Despite McGucken's caution, Marshall moved ahead with the suit, filing in the original jurisdiction of the California State Supreme Court.

In reply, the state's attorney issued a wide-ranging brief that defended the state's antimiscegenation statute from multiple perspectives, arguing that the state had a reasonable right to regulate interracial marriage. Marshall countered, bringing to bear an abundance of the available judicial, biological, and sociological evidence to refute the state's claims. Drawing on a variety of court rulings, Marshall argued that mixed marriage did not represent a sufficiently clear and present danger to the state as to justify the abrogation of the natural right of persons to marriage or to the free exercise of religion. Since the issue had never come before the California Supreme Court or the United States Supreme Court, the state's brief pointed to several other state and federal rulings. In analyzing the cases cited by the state, Marshall argued that those decisions represented the codification of racial prejudice and unfounded assumptions about "inferiority." Marshall skillfully pointed to the ways in which these rulings, and the state's use of them, worked only to uphold the ideology of white supremacy. He included an extended quotation from the ruling of an 1890 federal case in Georgia to make his point: "The amalgamation of the races is not only unnatural, but it is always productive of deplorable results. . . . [Mixed marriages] are productive of evil, and evil only, without any corresponding good."[50] Marshall followed this racist rhetoric with an extended quote from *Mein Kampf* that mirrored much of the language available in the rulings from the state and federal courts. Explaining the parallel, he charged California with pursuing the purity of the "blood of the so-called white race" at all costs.[51]

Marshall then moved on to answer the state's claims of biological justification for the statute. These claims were firmly rooted in the eugenics literature of the 1910s and 1920s. The state turned to Davenport's study, *Race Crossing in Jamaica*, and to the work of the Harvard geneticist W. E. Castle, that of University of California biologist S. J. Holmes, and of University of Chicago sociologist E. B. Reuter, for evidence of the deterioration that would result from interracial marriage. In each case, Marshall was able to point to analysis by Otto Klineberg and Ashley Montagu that illuminated the racist assumptions, methodological flaws, and the thin data in the early work. Klineberg and Montagu stood among the younger generation of social scientists who had recently put forth a potent argument for the cultural and sociological constructs of racial difference.[52] As Elazar Barkan has argued in *The Retreat of Scientific Racism*, this shift in perspective—from viewing race as a biological reality to viewing race as a social construction—

spelled the demise of eugenics among scientists in the late 1930s and early 1940s.[53] Marshall's need to refute the state's claims serves as a reminder that such perspectives lingered in the areas of law and policy long after they were discredited in the halls of academe.

Finally, Marshall turned to the sociological concerns raised in the state's brief. In his response to Marshall's initial filing, the state attorney used John LaFarge's *The Race Question and the Negro* to refute Marshall's claims about church teaching on interracial marriage. In that text, LaFarge counseled: "Where such intermarriages are prohibited by law, as they are in several States of the Union, the Church bids her ministers to respect these laws, and to do all that is in their power to dissuade persons from entering into such unions."[54] Far from a call for civil disobedience, LaFarge's message suggested that the disparity of conditions occasioned by racial difference was so great, and the danger to the unity of the marriage bond so severe, that an interracial union was simply not "prudent." Since social and cultural conditions changed ever so slowly, people had little choice but to accept them in the meantime. From this perspective, LaFarge offered a twist on the new paradigm of racial thinking: though he and others now acknowledged that racial categories had no objective biological foundation, the prejudices that had accrued along with this false biological argument simply had to be waited out.[55]

Though Marshall did not believe that the cited passage counseling priests to respect existing laws had any bearing on his claims, he requested LaFarge's comment.[56] As a result, LaFarge had a chance to clarify a position in private correspondence that he did not explicitly state in his public writings: "Respecting the laws does not mean that one approves of the laws or considers them either just or equitable."[57] LaFarge told Marshall that if the parties to the marriage had fully considered the social consequences of their marriage (presumably the difficulty they would have finding an accepting community in which to live and raise their children), then the Catholic Church would have no grounds on which to object. "Since the exercise of prudence is something which falls entirely within the competence of the contracting parties, it is altogether improper and immoral for the State to lay down a regulation upon a matter over which it has no competence. While prudence may be dictated to individuals as the more desirable course, that of complying with an unjust law under certain circumstances, changed circumstances such as the world is now engaged in would seem to make it equally the part of prudence to see that such laws are done away with and to register a protest against them."[58] In this way, LaFarge expressed his support for Marshall's venture—support that he offered again in subsequent correspondence.[59] As a result, Marshall was not discouraged by the state's use of LaFarge's published work.

Based on his correspondence with LaFarge, Marshall dismissed the state attorney's use of the text from *The Race Question and the Negro* by arguing that LaFarge objected to interracial marriages in particular cases but that the statute in question barred them completely. Also, Marshall claimed that the "social tensions" with which the Jesuit was concerned were outside of the ability of the state to legislate. Marshall turned to analogy to undermine the notion that social tensions were sufficient reason to legislate against interracial marriage: "The wedding of May and December, within the age limits of the statute, of the cultured to the ignorant, of the sick to the strong, of the poor to the rich, of the handsome to the ugly, of the Jew to the Gentile, of the Protestant to the Catholic, in none of these does the state venture to express a judgment."[60] Thus, Marshall returned race to the realm of social and cultural difference, highlighting the goals of white supremacy served by the antimiscegenation statute.

In October 1948, the California supreme court justices took the momentous step of being the first high court in the United States to strike down an antimiscegenation statute. The decision was close, with a majority of four justices and a minority of three. The dissent, written by Justice John Shenk, restated in great detail the points and arguments of the state's case, but it was the majority opinion that broke new legal ground.[61] Justice Roger Traynor's opinion shows clear signs that he had absorbed some of Daniel Marshall's reasoning. Though there was a definite consideration of the free exercise of religion question, that angle was subsumed by the demands of the Fourteenth Amendment. Thus, in summarizing the points at stake, Traynor wrote: "If the miscegenation law under attack in the present proceeding is directed at a social evil and employs a reasonable means to prevent that evil, it is valid regardless of its incidental effect upon the conduct of particular religious groups. If, on the other hand, the law is discriminatory and irrational, it unconstitutionally restricts not only religious liberty but the liberty to marry as well."[62] Traynor rehearsed the case law establishing marriage as a fundamental right that includes the right to marry the person of one's choosing. The California antimiscegenation statute, he said, restricted that choice by placing whole classes of persons out of bounds without establishing that there was clear reason for the restriction. He also rejected the arguments of both physical inferiority and social tension as a reason for the restriction. Citing cases related to jury selection and residential segregation, Traynor concluded that promoting peace could not come at the expense of fundamental constitutional rights. Ultimately, Traynor laid the blame for social tensions on racial prejudice. If the progeny of mixed marriages do indeed suffer the stigma of inferiority and rejection by both races, he argued, then "the fault lies not with their parents, but with the prejudices in the community and the laws that perpetuate those prejudices by giving legal force to the belief that certain races are inferior. If miscegenous marriages can be prohibited because of tensions suffered by the progeny, mixed religious unions could be prohibited on the same ground." In making this analogy, Traynor pointed directly to John La-Farge, noting that the Jesuit called the unions "not unlike" one another in the stress that such differences could place on a marriage. Finally, Traynor faulted the statute for failing to clearly define all of the terms employed to designate racial groups.[63]

The concurring opinions also expressed dismay at the arguments used by the state to justify the antimiscegenation law. Calling the statute a "product of ignorance, prejudice and intolerance," Justice Jesse Carter cited the Declaration of Independence, the Bill of Rights, the Fourteenth Amendment to the Constitution, and the Charter of the United Nations as important documents that the statute contradicted. He had been particularly moved by Marshall's use of the quotation from *Mein Kampf*, which he reproduced in full.[64] Justice Douglas Edmonds wrote a concurring opinion that came the closest to affirming Marshall's claims about free exercise. In large part, Edmonds agreed with Traynor, but he argued that freedom to marry was protected under the guarantee of religious freedom and that the proper emphasis should be placed on the absence of a clear and present danger that could justify compromising the First Amendment right of free exercise.[65]

Both Catholic and secular media swiftly reported on the *Perez* ruling, though they received it slightly differently. The secular media in particular recognized the willingness of some Catholics to challenge antimiscegenation statutes as evidence of the emerging presence of Catholics in the national civil rights movement. For instance, the *Nation* explained that "Marshall's achievement is a personal triumph, for most of the civil-rights organizations failed or refused to participate in the case on the assumption that miscegenation statutes could not be successfully challenged in the court."[66] Reports of the decision in Catholic journals tended to laud the *Perez* ruling as a moral victory while continuing to sound a note of caution about the dismal social and cultural environment for interracial marriages. The editors of *America* (led by John LaFarge, who served as associate editor, executive editor, and eventually editor-in-chief, until his death in 1963) quoted directly from Francis Connell's 1938 piece in the *AER* that called for Catholics to resist intrusive legislation. However, the quote from Connell was followed by a caution about the "great personal problems and difficulties" that an interracial marriage would entail. Gravely, the editors remarked: "There are few people who can accept such a burden. Toward them the attitude of Catholics will be dictated by respect for the person redeemed by Christ and the sacrament instituted by Him."[67] The tone of such a commentary suggests that the editors were far from ready to encourage interracial couples, even as they counseled respect for their decision to enter into the institution of marriage.

Some months later, *America* carried an article written by the Jesuit legal scholar Robert F. Drinan. Even though the article was entitled "Triumph over Racism," Drinan did not share the optimism of the writers at *The Nation*, as evidenced by his final word of advice: "There should, of course, be no agitation to repeal such statutes since 1) it is unrealistic to expect any such repeal, and 2) such a course of action might perpetrate the fallacy that Negroes, as a general practice, desire to intermarry."[68] With this statement he replicated the reluctance of Los Angeles's Auxiliary Bishop McGucken and many members of the hierarchy to risk challenging traditional taboos against interracial relationships in the name of racial justice. This final caution received a stinging rebuke from Ted LeBerthon, a Catholic journalist from Los Angeles, who suggested that Drinan had been struck by "an attack of excessive prudence" and that "the Church, in the interest of true prudence, should encourage the repeal of laws everywhere against interracial marriage, against anything that would intimate that our brother in Christ, the Negro, is something less than a human person."[69]

The coverage in the *Interracial Review* was much more positive than that in *America*, but the editors still felt compelled to strike a balance between reticence and the church's teachings. While they took into account LaFarge's cautions on social conditions and the state's use of them, the editors cast the problem of contemporary social circumstances as arising "not from anything in their marriage itself, but from the attitudes of the group around them"—attitudes that they were working to change. They argued that LaFarge's position helped to clarify church doctrine that declared the "absurdity and wrongness of any regulations which would take this matter out of where it properly and alone belongs: the free choice of the individuals concerned." This reading of LaFarge's position cast the problem of social conditions as something that could be dealt with through activism in the pursuit of racial justice. The editors lauded the California decision for correcting a "major moral and legal" scandal, and "because it is a powerful exemplification, in an unexpected quarter, of the far-reaching bearings of Catholic moral and sacramental teaching upon human conduct: of the Church's power to heal a wound that the accumulated wrongs of centuries have inflicted upon American society."[70]

The editors of the *Interracial Review* recognized the potential contained within Marshall's willingness to place canon law in conflict with the civil code of law. Marshall's argument in the *Perez* case illuminates the ways in which religious institutions and social teachings have provided alternative narratives that work to resist oppression. Catholics like Marshall were willing to creatively mine their traditions and teachings for new ways to pursue social reform. In that vein, Catholics continued the process of negotiating their public identity as both Catholics and Americans while they challenged the coercion and ideology of racist marriage legislation with their own sacred code of laws. Differing opinions among Catholics make it clear that Marshall's perspective in 1948 was far from the norm; many Catholic thinkers argued for a safer route, which amounted to tacit collaboration with discrimination. Yet Marshall and other activists like him insisted that there was an alternative path. In very publicly pressing the case against eugenicists' efforts to shape marriage, they would expand our contemporary understanding of the civil rights movement and force us to acknowledge a wider significance for Catholic activism. The social struggles of Catholics like Marshall would only grow, quietly but steadfastly, in the coming decades.[71]

By 1948, when Daniel Marshall successfully challenged California's antimiscegenation law, the eugenics movement was in full-scale retreat from the positions of its heyday in the 1920s. The scientific community, led by developments in biology and anthropology, now dismissed the simplistic, racist, and prejudicial eugenic philosophy of heredity. In doing so, they joined the chorus of Catholic opponents who had been working against negative eugenic measures for decades.

While Catholics remained at odds with those who supported negative eugenic measures, their long-standing pronatalism dovetailed, at times a bit too easily, with the shift of the "new eugenics" toward a more positive tone. Encouraging strong and fertile marriages appealed both to eugenicists and to Catholics, and as the 1940s wore on, the first fruits of the baby boom seemed to confirm both groups' enthusiasm. Nonetheless, eugenicists were selective in their pronatalism, making a clear distinction between those who should be encouraged to reproduce and those who should be counseled to make effective use of birth control or to undergo voluntary sterilization. This selectivity guaranteed that eugenicists and Catholics would never fully find common ground.

Catholic activists continued to reject the efforts of state legislatures to determine who was fit to marry. Moral theologians clearly refuted the right of the state to impose health restrictions and requirements on marriage. They maintained that the church had sole jurisdiction over the marriage contract and held the authority to decide which parties were eligible to marry. While Daniel Marshall was the first Catholic (and one of the first Americans) willing to argue against such laws in court, he had based his argument on the centuries-old central tenets of Catholic theology about marriage and family, seeing them to their logical conclusion. In this, Marshall joined the many clergy and laypersons over the decades who had stood up to object to eugenic measures that rendered unto the state the power to decide who was allowed to bear children and who was not. In the face of the "new eugenics," which offered a much more subtle approach to racial hierarchy and efforts to curb the reproduction of the "unfit," such willingness to challenge the status quo would continue to be necessary to guarantee the respect for individual rights and bodily integrity that had animated Catholic opposition to eugenics in the decades prior to World War II.

Epilogue

In the fall of 1953, Pope Pius XII met with a large group of distinguished geneticists from around the world. The gathering at his summer home, Castel Gandolfo, gave the Holy Father a chance to address the church's attitude toward genetics in the wake of World War II. In doing so, he informed the scientists that there was nothing in Catholic teaching that represented a barrier "to research, to knowledge, to affirmation of truth." Rather, Catholic teaching served to balance sources of authority and to encourage researchers to check one hypothesis against other possible explanations. He affirmed his support for the goals of genetics, albeit in his own very particular, and proactive, definition of the field: "to influence the transmission of hereditary factors in such a way as to promote what is good and to eliminate what is harmful." In this, Pius XII joined with the eugenicists in affirming that science could be applied to improve social welfare, but there was no hint that he affirmed their goals or methods. As with scores of moral theologians before him, Pius XII took care to express his disapproval for some of the means that scientists and politicians employed to achieve those goals, especially singling out Nazi racial ideology, marriage restriction, and eugenic sterilization. He particularly sought to call attention to the continued use of eugenic sterilization. Despite the defeat of the Germans in World War II and the end of Nazi racial policies, he noted, the medical profession, social workers, and other activists continued to promote sterilization as a way to prevent the transmission of hereditary disease, albeit not in the context of broad state-sponsored programs. In this address to the geneticists, Pius XII basically reiterated the importance of the teaching on eugenics that Pius XI had articulated in *Casti Connubii*.[1] In the twenty-three years since Pius XI's encyclical, much of the world had changed. The Catholic perspective on negative eugenics, however, had not.

The church's steadfast opposition to negative eugenic measures was an increasingly rare instance of unwavering moral principles in a world that, at least in the eyes of Pius XII, was becoming ever more relative.

Despite the consistency demonstrated by the pope's address at Gandolfo, the 1950s would continue the radical shifts of the previous two decades, particularly in science and anthropology, and in the place of Catholics within American society. For geneticists, the landscape of inquiry was fundamentally altered in the spring of 1953 when James Watson and Francis Crick announced that they had discovered the structure of deoxyribonucleic acid.[2] Their discovery, combined with the demonstration in 1944 that DNA carried the markers of biological specificity in bacteria, was the key that unlocked the new world of molecular biological, soon to upend the way that science understood the issue of heredity.[3]

Similarly, by the end of World War II, as we have seen, culture had become the predominant factor in social scientific thinking about racial difference. As support for the science of eugenics declined, prominent anthropologists and biologists, such as Franz Boas, Ashley Montagu, Ruth Benedict, and Julian Huxley, began to transform the way that people explained human diversity, and to thoroughly undermine the ideas about biological racial distinctions that had formed the underlying principles of eugenics. Rejecting the notion of strict biological differences between racial groups, anthropologists now focused on environment, culture, and human changeability as the way to understand human difference. Thus, a movement in anthropology that started in the 1930s began to gain wider public support. Huxley's *We Europeans* (1935), Montagu's *Man's Most Dangerous Myth* (1942), and Ruth Benedict's *The Races of Mankind* (1943) expanded upon the work of Boas and created a corpus of material that rejected biological notions of race. The movement away from biological notions of race was so strong that in 1950 UNESCO published *The Race Concept*, a collection of statements from all corners of the global scientific community. The text argued for the essential unity of the human race across three primary groupings (Caucasian, Negroid, and Mongoloid) and made several essential points: that the differences which do exist between human groups are the result of both heredity and environment; that cultural traits and racial traits are not connected; that cultural traits cannot be explained by genetics; that intelligence is not a racial characteristic; and that racial purity is a myth.[4] These ideas shaped not just scientific understanding. By undercutting eugenicists' claims, they also changed cultural expectations and political strategies.

The shifting ideology of culture and race must have influenced Will Herberg when he set out in the mid-1950s to write his sociological analysis of religion in the United States. In *Protestant—Catholic—Jew*, Herberg wrote, "For all its wide variety of regional, ethnic, and other differences, America today may be conceived, as it is indeed conceived by most Americans, as one great community divided into three big sub-communities religiously defined, all equally American in their identification with the 'American Way of Life.'" Additionally, Herberg argued that Protestantism, Catholicism, and Judaism were essentially all versions of a larger "American religion" that shared a common set of spiritual values that affirmed American democracy. As a result, Herberg maintained, "By the second quarter of the present century the American Catholic, like every other American, was thinking of his church as one of the three 'religions of democracy,' side by side with the other two."[5]

This sense that Catholicism was compatible with—and indeed, inextricable from—American political culture was a reflection of the views held by Catholics themselves. The decades-old migration of US Catholics away from the isolation of their earlier, insulated communities continued with even greater speed in the postwar years. A number of scholars of US Catholicism have remarked on the key factors that contributed to Catholics' integration into mainstream American culture during the course of the twentieth century. First, the GI Bill allowed tremendous numbers of Catholics to partake of higher education in both Catholic and secular universities for the first time. These college experiences gave Catholic veterans the tools to join America's professional class and to achieve middle-class status on an unprecedented scale. Second, the embrace of anti-Communism by prominent Catholics, and gradually by Catholics in general, helped to integrate them into the liberal consensus culture of the Cold War. In that cultural environment a few Catholics, like Bishop Fulton Sheen and the Trappist monk Thomas Merton, became full-fledged media personalities, helping to bring a benign image of Catholicism to the non-Catholic public. Finally, the increase in birthrates nationally after the war made the traditional Catholic pronatalism seem like a variation on the norm rather than evidence of a vastly different—and, at least according to some, a suspect—approach to family and parenting. Thus, Catholics in the 1950s maintained their religious traditions and community but also managed to integrate successfully with the larger American population in most respects.[6] A snapshot of the Catholic population and their situation in the American cultural context in the years after World War II would hardly bear the traces of the years of conflict between Catholics and eugenics advocates over both ideas and policy initiatives. The eugenics movement had withered away, and by 1960 there was a Catholic in the White House.

That is not to say that eugenically motivated policies totally disappeared from American life after World War II. Eugenic sterilization policies have never been found unconstitutional, though many states have repealed their laws. While forced sterilization waned in the 1940s in most locations, there were some conspicuous exceptions. Advocates in Alabama continued to attempt to pass a law authorizing the sterilization of institutionalized person. Also, in areas where eugenic sterilization programs continued, such as North Carolina, women could find themselves caught up in a variety of conflicting and coercive situations, depending on whether they were being forced into sterilization or whether they sought the operation as a form of birth control.[7] Indeed, the emergence of a voluntary sterilization movement in the postwar period—shed of its explicit eugenic origins but still tied to population control and concerns about the reproductive potential of the developmentally disabled—and the laws that regulated access to the operation, added significant complexity to both the moral and public policy questions associated with the practice.[8]

For most of the nation, forced sterilization did not arise again as an issue until the 1960s and 1970s, when poor people of color were targeted for the operation. A disproportionate number of black women were sterilized under federally funded family planning programs, and Indian Health Service hospitals performed thousands of sterilizations without informed consent. The abuses led the Department of Health, Education, and Welfare in 1973 to issue guidelines for sterilizations performed at federally funded clinics. Those guidelines were challenged in court, where they were struck down in 1974 on the basis that HEW did not have the resources to adequately monitor the individual consent they now required for sterilization operations. The litigation resulted in revised guidelines in 1978 that prohibited federal funding for the sterilization of persons under the age of twenty-one, persons who were mentally incompetent, and persons who were institutionalized. Anyone else who wished to undergo sterilization was required to sign a special consent form and undergo a thirty-day waiting period. The guidelines also specified the range of education and information counselors needed to provide for patients.[9] This tightening of the guidelines helped to reduce instances of abuse, at least where federal funding was concerned.

Today, compulsory sterilization is unheard of, but we continue to struggle with the historical legacy of the widespread state programs. North Carolina's legacy goes back much further than that of many other states. Forced sterilization there ceased only in 1977. Since 2002, the state has been in the midst of a wide-ranging public discussion about its past, even going so far as to issue an official apology. Recently, a task force appointed by Governor Bev Perdue has been mulling over what the state might owe to the thousands of victims who were robbed of their childbearing capacity without their consent. The prospect of paying tens of thousands of dollars in reparations for the roughly 3,000 remaining victims from North Carolina gives just a glimpse of the larger debt owed to the upwards of 60,000 individuals sterilized over the course of the twentieth century.[10]

Despite the cultural and political achievement many Catholics felt at becoming comfortably part of the Cold War consensus culture, their commitment to the principles that were the foundation of their objections to negative eugenics initiatives continued to set them apart. By the mid-1970s, the public face of Catholic teaching with regard to reproductive issues focused not on eugenic policy initiatives but on the controversy over contraception and a vociferous response to the legalization of abortion in the *Roe v. Wade* decision. Though eugenics was not the primary issue, it still lurked in the background. In 1975, the Vatican yet again affirmed that sterilization remained "absolutely forbidden according to the doctrine of the church, notwithstanding whatever good intentions of the person acting to provide the cure or the prevention of a physical or psychic evil which is foreseen or is feared would derive from pregnancy." The statement reaffirmed that sterilization is "harmful to the dignity and inviolability of the human person."[11] Official Catholic teaching on the issue was nearly identical to the positions developed in the early decades of the twentieth century.

When initially faced with the question of sterilization for criminals in 1910, theologians acknowledged that if there were no other alternatives to protect the common good, the operation might be permissible. But there were other alternatives, and the danger of public authorities abusing their power was significant. From the beginning of the eugenics movement, Catholics writers and thinkers seemed intuitively aware of the broad constellation of issues beneath the seemingly straightforward desires of eugenicists. Calls for intelligence testing, immigration restrictions, and antimiscegenation laws were based on an understanding of human worth that privileged racial hierarchy and discriminated against whole ethnic groups, the impoverished, and those whose behavior failed to conform to middle-class norms. Catholics involved with the eugenics movement, such as John Ryan and John Montgomery Cooper, supported the idea that society should seek to improve the general health and welfare of the population, but they were adamant that this improvement could not come for some at the expense of others. At the news of the Supreme Court's ruling in the *Buck v. Bell* case, the editors of *America* made the statement from which the title of this book is drawn, that every person is "an image of God" and "not a mere social factor."[12]

This distinction between those who privileged a hierarchy of human worth and those who took a universalist approach to humanity is one of the keys to understanding the Catholic response to eugenic ideas and policies. For Catholics, the moral theology undergirding their opposition to negative eugenics policies was primary. Along with their moral objection to the discrimination implied in eugenic policy initiatives, Catholic thinkers, writers, and activists also clarified their position on the appropriate role of the state. They argued that each individual had certain natural rights that the state could not infringe upon. Thus, they defended Pius XI's stance in *Casti Connubii* that "public magistrates have no direct power over the bodies of their subjects."[13] The notion that the integrity of the human body is beyond the reach of the state was matched by the equal insistence on maintaining the church's sphere of authority. Catholics argued that marriage was fundamentally a sacrament, and that the church had the sole jurisdiction over the ability to sanction or impede it.

In the process of defending these positions, members of the clergy, church hierarchy, and laity developed an effective political strategy that showed an awareness of the features and constraints of American pluralism. As the twentieth century wore on, Catholics experimented with increasingly savvy techniques for getting their message across. In the 1920s, for example, they played down their religious beliefs in order to more effectively make their case with legislators, who might otherwise have been wary of their "otherness." Recognizing that they lived in a society that demanded a separation between church and state, Catholics developed a growing array of arguments against such measures, drawing on the latest developments in biology, medicine, psychology, sociology, and the law to convince non-Catholics that rejecting eugenics policies was in the best interest of the entire community.

We must acknowledge, as well, the moments of uncomfortable overlap between Catholics and eugenicists. The Catholic pursuit of the common good has been far from smooth, nor devoid of the social and cultural biases that can infect us all. While certain instances of Catholic support for, and cooperation with, positive eugenics efforts fit within the teachings of the church, other moments are less salutary. The hesitancy of the Catholic hierarchy to challenge antimiscegenation statutes after World War II, for example, reminds us that Catholics are as susceptible to the pull of societal inertia as anyone else. We all can be lulled into choosing the easier path, into choosing silence over action, gradualism and the status quo over justice.

Yet these moments of Catholic capitulation are themselves instructive. They remind us that social and political change nearly always requires the bravery of the few, of men and women like Mary Hawks, who first spoke out against sterilization legislation in New Jersey, or Kathryn McCarthy, who raised the alarm about the outrageous sterilizations of girls in Kansas, or Daniel Marshall, who defeated California's antimiscegenation statute in court. In contemporary America they are recognizable to us, both as progenitors of the movements for reproductive and racial justice and as the embodiment of a fundamental American archetype—the individuals who struggle against the unfairness of their age, who reveal the country's complicity in racism and sexism and bias. Such figures dot our country's historical landscape and too frequently their names are largely unknown.

Not as familiar is the place of professing Catholics themselves within that landscape of social change. It is easy, particularly in our pessimistic political moment, to doubt the value of huge, lumbering institutions and the ability of their devotees to advocate for positive change. And what better example than the Catholic Church! As an age-old institution, its faults are many, but the rise and fall of eugenics in America also forces a different perspective. The church is its members, and their objections to negative eugenics policies were built on moral teaching that remained essentially unchanged across the twentieth century. While there are many things that we may like to change about the Catholic Church, we must also acknowledge moments like these, when its institutional steadfastness was a source of astounding strength and inspiration for clergy, laypersons, and non-Catholics dedicated to human rights. The dogged and sustained activism of Catholics made them the primary source of opposition to eugenic sterilization and helped to influence the rejection of negative eugenics initiatives even in places where Catholics did not dominate the public discourse. And at the center of this half-century struggle with eugenics was an unchanging vision of the primacy of all human life.

ACKNOWLEDGMENTS

During the course of my work on this book, I have accrued many debts—happily. Some of those debts are institutional on their face, but in fact, each one of them represents a set of personal relationships without which I would not have been able to complete this project. In the search for primary source material, I have been guided by a number of kind and knowledgeable archivists. I owe special thanks to Timothy Meagher and W. John Shepard at the Archives of the Catholic University of America; Robert S. Cox at the American Philosophical Society; and Chris Krosel at the Archives of the Diocese of Cleveland. My time at the American Philosophical Society in Philadelphia was made possible through a generous residential fellowship from the Andrew W. Mellon Foundation. Additional research trips were funded by the American Studies Department at the University of Minnesota. Dissertation fellowships from the University of Minnesota Graduate School and the Institute for the Advanced Study of Religion at Yale University supported my research and writing in the early stages of this work.

In addition to this institutional support, I have had the good fortune to encounter many wise scholars who encouraged my work along the way. R. Scott Appleby, Jon Butler, Peggy Pascoe, Peter Steinfels, and Leslie Woodcock Tentler granted me their time and advice through the years. Segments of the book are exponentially better due to the commentary and conversation stimulated by sharing my work at meetings of the American Studies Association, the American Society of Church History, and the American Academy of Religion. In particular, I would like to thank Sandra Yocum-Mize and Amy Laura Hall for their perspective and encouragement.

The editorial process of review and response is essential to producing fully realized scholarship, and I have benefited from that process at many stages in the development of this work. Portions of chapters 2, 3, and 6 have appeared as journal articles (see Leon, "'A Human Being, and Not a Mere Social Factor': Catholic Strategies for Dealing with Sterilization Statutes in the 1920s," *Church History* 73, no. 2 [June 2004]: 383–411; Leon, "'Hopelessly Entangled in Nordic Pre-suppositions': Catholic Participation in the American Eugenics Society in the 1920s," *Journal of the History of Medicine and Allied Sciences* 59, no. 1 [January 2004]: 3–49; and Leon, "Tensions Not Unlike that Produced by a Mixed Marriage: Daniel Marshall and Catholic Challenges to Interracial Anti-miscegenation Statutes," *U.S. Catholic Historian* 26, no. 1 [December 2008]: 27–44). The thoughtful criticism of the editors and the reviewers at *Church History*, the *Journal of the History of Medicine and Allied Sciences*, and *US Catholic Historian* made those texts much stronger and began the transformation of the original manuscript into its current state. Similarly, my anonymous reviewers for the full manuscript contributed important critiques and insight that helped me refine the manuscript. Doug Mitchell and Tim McGovern at the University of Chicago Press have patiently shepherded me through the process of review and publication. Finally, in the late stages of revision, David Lobenstine's editorial guidance proved an invaluable help in making this a much better book.

I have also had the good fortune to inhabit several communities of scholars over the years that I have worked on this research and writing. There were a host people at the University of Minnesota whose intelligence and support shaped my work and thinking; Elaine Tyler May and Riv-Ellen Prell have done the most strenuous work as my trusted advisors. I will forever be in their debt for the lessons I continue to learn from them about the life of the mind, collegiality, and kindness. David Noble has served as an intellectual father and a cherished friend. Additionally, the companionship and support of my extended cohort and the advisors in American Studies helped ease me through my years in graduate school. I would like to thank Hokulani Aikau, Matt Basso, Matt Becker, Karla Erickson, Sara Evans, David Gray, Colleen Hennen, Kate Kane, Sally Gregory Kohlstedt, Joyce Mariano, Lary May, Mary Rizzo, David Roediger, Felicity Schaeffer, Amy Tyson, and Rudolph Vecoli. After I left Minnesota, I landed in the warm and generous environment of the Roy Rosenzweig Center for History and New Media in the Department of History and Art History at George Mason University. Roy Rosenzweig was a trusted and generous colleague who supported me through the revision process. I miss his kindness and advice every day, but his wonderful spirit lives on in the work of the people who carry out the mission of the center and who have been important to the later stages of this work: Sheila Brennan, Dan Cohen, Christopher Hamner, Patrice Mortson, Tom Scheinfeldt, Sean Takats, and Sarah Richardson Whelan. I have also benefited from the wisdom of colleagues and friends in American Studies at Georgetown University: Randy Bass, Hugh Cloke, Elizabeth McKeown, Diana Owen, and my students, who teach me new things every day. I thank them all for their energy, innovation, and encouragement.

Finally, my family has been incredibly supportive of my work. Michael McQuillen, in his many roles as neurologist, bioethicist, and uncle, has lent his expertise, his books, and his skills as a reader all along the way. During most of my research trips, James Leon's house was a refuge. He always provided me with a warm meal and good dose of big-brother love and support. Finally, my mother, Mary Devlin—who I am sure now knows far more about Catholic archival sources and the history of eugenics than she could have imagined years ago—has generously listened as I painstakingly narrated every instance of writer's block and every breakthrough. I cannot remember a time when I could not depend upon her confidence in me to sustain me, even when my own confidence faltered. I owe each of them my love and my thanks.

NOTES

NOTES TO A NOTE ON SOURCES

1. Catholic Library Association, *The Catholic Periodical Index: A Guide to Catholic Magazines* (Scranton, PA: National Catholic Educational Association, 1930–).

NOTES TO INTRODUCTION

1. Oliver Wendell Holmes Jr., *Majority Opinion,* Buck v. Bell, 274 US 200, 47 S. Ct. 584 (1927).
2. "Unjustified Sterilization," *America,* 14 May 1927.
3. Elazar Barkan, *The Retreat of Scientific Racism: Changing Concepts of Race in Britain and the United States between the World Wars* (Cambridge: Cambridge University Press, 1992).
4. Carl Degler, *In Search of Human Nature: The Decline and Revival of Darwinism in American Social Thought* (New York: Oxford University Press, 1991), 145–51, 179–86.
5. Dennis L. Durst, "'No Legacy Annuls Heredity from God': Evangelical Social Reformers and the North American Eugenics Movement" (PhD diss., Saint Louis University, 2002).
6. In using the phrase "re-imagining" I am signaling my debt to the work of Benedict Anderson on nationalism (see his *Imagined Communities: Reflections on the Origins and Spread of Nationalism,* rev. ed. [London: Verso, 1991]). Anderson defines a nation as "an imagined political community . . . both inherently limited and sovereign" (6). Eugenicists used a variety of means, particularly widespread access to inexpensive printed publications, like newspapers, to "imagine" a racially superior national community—one that was free from biological degeneracy and all the attendant social ills that they thought accompanied it. The result would be a community with secure borders that contained a citizenry composed of superior individuals who could rightfully govern themselves and safeguard the future of the nation.
7. Quoted in Daniel J. Kevles, *In the Name of Eugenics: Genetics and the Uses of Human Heredity* (Cambridge, MA: Harvard University Press, 1995), 63.
8. "Report of Sub-Committee on the Ultimate Program to be Developed by the Eugenics Society of the United States of America," *Eugenical News* 8, no. 8 (August 1923), 73.
9. Ibid., 79.

10. Charles B. Davenport, *Heredity in Relation to Eugenics* (New York: Henry Holt, 1911), 4.
11. Michael Omi and Howard Winant, *Racial Formation in the United States: From the 1960s to the 1990s*, 2nd ed. (New York: Routledge, 1994), 56.
12. Diane B. Paul, *Controlling Human Heredity, 1865 to the Present* (Amherst, NY: Humanity Books, 1998), 17–19; and Edward J. Larson, *Sex, Race, and Science: Eugenics in the Deep South* (Baltimore: Johns Hopkins University Press, 1995), 2.
13. Matthew Frye Jacobson, *Whiteness of a Different Color: European Immigrants and the Alchemy of Race* (Cambridge, MA: Harvard University Press, 1998), 1–14, 39–90 (quotation, 7).
14. Ian Haney López, *White by Law: The Legal Construction of Race* (New York: New York University Press, 1996), 10.
15. Kevles, *In the Name of Eugenics*, 60–62.
16. Leslie Woodcock Tentler, "On the Margins: The State of American Catholic History," *American Quarterly* 45, no. 1 (March 1993).

NOTES TO CHAPTER 1

1. Stephen M. Donovan, "Circa Liceitatem Cujusdam Operationis Chirurgoae Proponuntur Dubia Nonnulla," *American Ecclesiastical Review* 42, no. 9 (March 1910). The initial question is posed in Latin.
2. Philip R. Reilly, *The Surgical Solution: A History of Involuntary Sterilization in the United States* (Baltimore: Johns Hopkins University Press, 1991), 30–33.
3. Ibid., 45–46.
4. Ibid., 34–39.
5. "The Morality and Lawfulness of Vasectomy," *American Ecclesiastical Review* 44, no. 11 (May 1911) (quotation, 563).
6. Ibid., 563–65. This notion—that the reproductive rights of criminals could be circumscribed—resurfaces in the sections of the encyclical *Casti Connubii* that deal with eugenic sterilization.
7. "Schmitt's Argument," *American Ecclesiastical Review* 44, no. 11 (May 1911) (quotation, 569).
8. Ibid., 570. Schmitt's reflections drew responses both from Stephen Donovan and also from Father Theodore M. Labouré, another theologian who had participated in the initial *AER* exchange. They both redirected the discussion away from the intention of the surgeon and the double effect of the operation and toward the question of whether or not the state had the power to order the operation to preserve the welfare of the common good (see Stephen M. Donovan, "The Morality of the Operation of Vasectomy," *American Ecclesiastical Review* 44, no. 11 [May 1911]: 571–74; and Theodore Labouré, "Is Vasectomy Then Unlawful," *American Ecclesiastical Review* 44, no. 11 [May 1911]: 574–83).
9. Stephen M. Donovan, "Summing up the Discussion on Vasectomy," *American Ecclesiastical Review* 45, no. 3 (September 1911) (quotation, 318). The conversation continued for months, including an article from Dr. Austin O'Malley that clarified the physical aspects of the operation and offered his own reading of the morality question (see Austin O'Malley, "Vasectomy in Defectives," *American Ecclesiastical Review* 44, no. 12 [June 1911]: 684–705). See also "The Morality and Lawfulness of Vasectomy," *American Ecclesiastical Review* 45, no. 1 (July 1911); Theodore Labouré, "One Last Remark on Vasectomy," *American Ecclesiastical Review* 45, no. 3 (September 1911); and "P. Schmitt on the Gravity of the Mutilation Involved in Vasectomy," *American Ecclesiastical Review* 45, no. 3 (September 1911).

10. Francis Galton, *Hereditary Genius: An Inquiry into Its Laws and Consequences* (London, 1869); and Francis Galton, *Natural Inheritance* (London, 1889). For an account of Galton's influence on the early eugenics movement, see Kevles, *In the Name of Eugenics*, 3–19. Other general works on eugenics in the United States include Mark H. Haller, *Eugenics: Hereditarian Attitudes in American Thought* (New Brunswick, NJ: Rutgers University Press, 1963); and Paul, *Controlling Human Heredity, 1865 to the Present*. Excellent regional/local studies are Nancy L. Gallagher, *Breeding Better Vermonters: The Eugenics Project in the Green Mountain State* (Hanover, NH: University Press of New England, 1999); and Larson, *Sex, Race, and Science*. There are also a number of good works on feeble-mindedness, intelligence, and racial difference: Stephen Jay Gould, *The Mismeasure of Man* (New York: W. W. Norton, 1981); Steven Noll, *Feeble-Minded in Our Midst: Institutions for the Mentally Retarded in the South, 1900–1940* (Chapel Hill, NC: University of North Carolina Press, 1995); Steven Selden, *Inheriting Shame: The Story of Eugenics and Racism in America* (New York: Teachers College Press, 1999); and James W. Trent, *Inventing the Feeble Mind: A History of Mental Retardation in the United States* (Berkeley: University of California Press, 1994).
11. Kevles, *In the Name of Eugenics*, 41–44.
12. Haller, *Eugenics*, 63–66; and Kevles, *In the Name of Eugenics*, 51, 54–56.
13. Peter J. Bowler, *Darwinism* (New York: Twayne, 1993), 36–37.
14. Don O'Leary, *Roman Catholicism and Modern Science: A History* (New York: Continuum, 2006), 19–22. The work of two American authors contributed significantly to the popularization of this framework of combat: John William Draper, *History of the Conflict Between Religion and Science* (New York: D. Appleton and Co., 1897); and Andrew Dickson White, *A History of the Warfare of Science with Theology in Christendom* (New York: D. Appleton and Co., 1901).
15. David C. Lindberg and Ronald Numbers, eds., *God and Nature: Historical Essays on the Encounter between Christianity and Science* (Berkeley: University of California Press, 1986); and John Hedley Brooke and Geoffrey N. Cantor, *Reconstructing Nature: The Engagement of Science and Religion* (Oxford: Oxford University Press, 2000). Specifically addressing the Roman Catholic Church, O'Leary's work *Roman Catholicism and Modern Science* goes into great detail about the complex social and political considerations that shaped an often varied response from the institutional church to scientific innovation and discovery. Also, a recent examination in the Vatican Archives suggests that there was never any organized or official effort to respond to Darwinian evolution. Rather, the Vatican addressed the work of individual Catholic writers and scientists, and those responses were often more concerned with European politics and theological conflicts than they were with evolution as a challenge to church authority (see Mariano Artigas, Thomas F. Glick, and Rafael A. Martínez, *Negotiating Darwin: The Vatican Confronts Evolution, 1877–1902* [Baltimore: Johns Hopkins University Press, 2006]).
16. O'Leary, *Roman Catholicism and Modern Science*, 9–16, 45–56. It was in this context that in 1864 Pope Pius IX issued the *Syllabus of Errors*. Though the document was composed with mostly Italian circumstances and questions in mind, it was interpreted within and without the larger church as a rejection of Western democracy and scientific principles, as the Pope refused to "come to terms with progress, Liberalism and modern civilization" (51). In the wake of the *Syllabus of Errors*, in 1869 Pius IX called for a rare general council of Catholic bishops. The most notable outcome of the First Vatican Council, which Pius tightly controlled, was the doctrine of papal infallibility. Liberal bishops were marginalized during the proceedings, and

the resultant documents upheld the traditional teachings of the church in the face of scientific and philosophical challenges.

17. "Scholasticism," in *New Catholic Encyclopedia* (Washington, DC: Catholic University of America, 1967), 12:1155–56, 1165.
18. Leo XIII, *Aeterni Patris* (Encyclical of Pope Leo XIII on the Restoration of Christian Philosophy, 4 August 1879), par. 1, http://www.vatican.va/holy_father/leo_xiii/encyclicals/documents/hf_l-xiii_enc_04081879_aeterni-patris_en.html.
19. Ibid., par. 30.
20. O'Leary, *Roman Catholicism and Modern Science*, 97–100; and R. Scott Appleby, "Exposing Darwin's 'Hidden Agenda': Roman Catholic Responses to Evolution, 1875–1925," in *Disseminating Darwin: The Role of Place, Race, Religion and Gender*, ed. Ronald L. Numbers and John Stenhouse (Cambridge: Cambridge University Press, 1999), 173–208.
21. R. Scott Appleby, *"Church and Age Unite!": The Modernist Impulse in American Catholicism* (Notre Dame, IN: University of Notre Dame Press, 1992).
22. Pius X, *Pascendi Dominici Gregis* (Encyclical of Pope Pius X on the Doctrine of the Modernists, 8 September 1907), par. 48, http://www.vatican.va/holy_father/pius_x/encyclicals/documents/hf_p-x_enc_19070908_pascendi-dominici-gregis_en.html.
23. Thomas J. Gerrard, "Eugenics and Catholic Teaching," *Catholic World*, June 1912, 291.
24. Gould, *The Mismeasure of Man*, 158–74; Henry H. Goddard, *The Kallikak Family: A Study in the Heredity of Feeble-Mindedness* (New York: Macmillan, 1912); and Henry H. Goddard, *Feeble-Mindedness: Its Causes and Consequences* (New York: Macmillan, 1914).
25. Thomas Vernor Moore, "Types and Causes of Feeble-Mindedness," *Proceedings of the Fourth National Conference of Catholic Charities* (Washington, DC: National Conference of Catholic Charities, 1916), 237.
26. Ibid., 239. Moore stood out as a pioneering Catholic psychiatrist whose work with children balanced a Catholic social perspective with rigorous scientific activity. Studying under Rev. Edward Aloysius Pace, and later under Wilhelm Wundt, Moore cultivated his interest in experimental psychology, developing a focus on the mental health of children. Then, while in medical school at Johns Hopkins, Moore came under the influence of Adolph Meyer, an important American psychiatrist who integrated Darwinian theories into psychology, approached illness in a holistic way, and supported the mental hygiene movement (see Benedict Neenan, *Thomas Verner Moore: Psychiatrist, Educator and Monk* [New York: Paulist Press, 2000], chap. 2–4).
27. Dorothy Ross, *The Origins of American Social Science* (Cambridge: Cambridge University Press, 1992).
28. Dorothy M. Brown and Elizabeth McKeown, *The Poor Belong to Us: Catholic Charities and American Welfare* (Cambridge, MA: Harvard University Press, 1997), 62–77.
29. Henry Somerville, "Eugenics and the Feeble-Minded," *Catholic World*, May 1917, 212.
30. Goddard, *Feeble-Mindedness*.
31. Somerville, "Eugenics and the Feeble-Minded," 216–17.
32. Gerrard, "Eugenics and Catholic Teaching," 292–98 (quotations, 297, 298).
33. Jacobson, *Whiteness of a Different Color*, 78–81. In *Parish Boundaries: The Catholic Encounter with Race in the Twentieth-Century Urban North* (Chicago: University of Chicago Press, 1996), John McGreevy tells the story of Catholics' growing interaction with African Americans, arguing that "the primary 'race' problem for American Catholics before the 1940s was the physical and cultural integration of the various

Euro-American groups into the parishes and neighborhoods of the urban North, not conflicts between 'blacks' and 'whites'" (9). For a summary of Catholic immigration and diversity, see Jay P. Dolan, *The American Catholic Experience: A History from Colonial Times to the Present* (Notre Dame, IN: University of Notre Dame Press, 1992) 127–57.

34. John Higham, *Strangers in the Land: Patterns of American Nativism*, rev. ed. (New York: Atheneum, 1975). Higham points to anti-Catholicism as one of the three main strains of American nativism, along with antiradicalism and Anglo-Saxon racial nationalism.
35. John T. Noonan, *Contraception: A History of Its Treatment by the Catholic Theologians and Canonists* (1965; repr., New York: Mentor-Omega, New American Library, 1967), 461–70; Ellen Chesler, *Woman of Valor: Margaret Sanger and the Birth Control Movement in America* (New York: Anchor Books, 1992), 34–38. On the Comstock laws, see also Andrea Tone, *Devices and Desires: A History of Contraceptives in America* (New York: Hill and Wang, 2001), 3–24.
36. Mission priests, on the other hand, who were only occasional visitors to established congregations, took a more direct approach to the issue, making specific inquiries and admonitions (see Leslie Woodcock Tentler, *Catholics and Contraception: An American History* [Ithaca: Cornell University Press, 2004], 15–42).
37. Chesler, *Woman of Valor*, 56–104.
38. Margaret Sanger, *Family Limitation*, rev. 5th ed. (New York: Margaret Sanger, 1916), 2–3.
39. John A. Ryan, "Family Limitation," in *Family Limitation and The Church and Birth Control* (New York: Paulist Press, 1916), 5.
40. Edward A. Ross, "The Causes of Race Superiority," *Annals of the American Academy of Political and Social Science* 18 (July 1901). For a treatment of race suicide, see Gail Bederman, *Manliness and Civilization: A Cultural History of Gender and Race in the United States, 1880–1917* (Chicago: University of Chicago Press, 1995), 200–206; Thomas G. Dyer, *Theodore Roosevelt and the Idea of Race* (Baton Rouge: Louisiana State University Press, 1980), 143–67; Linda Gordon, *Woman's Body, Woman's Right: Birth Control in America*, rev. ed. (New York: Penguin, 1990), 137–41; and Miriam King and Steven Ruggles, "American Immigration, Fertility, and Race Suicide at the Turn of the Century," *Journal of Interdisciplinary History* 22, no. 3 (Winter 1990).
41. Ryan, "Family Limitation," 12; Jacobson, *Whiteness of a Different Color*, 52.
42. Ryan, "Family Limitation," 9–10.
43. Ibid., 12.
44. Ibid., 13–16.
45. Born and raised in Minnesota, John Ryan grew up under the influence of the labor politics of the Knights of Labor and Edward McGlynn, but also informed by church teaching. As a result, Ryan bridged the gap between Catholic thought on social questions and contemporary reform. In pursuit of his degree in Sacred Theology at the Catholic University of America in 1906, he produced a dissertation that echoed Pope Leo XIII's sentiments on the rights of labor in his encyclical *Rerum Novarum* (1893) and that would become the foundational text for twentieth-century Catholic engagement with labor issues. In *A Living Wage: Its Ethical and Economic Aspects* (New York: Macmillan, 1906), Ryan argued that wages should not be treated as a bargaining issue between worker and employer, but that each workingman had a natural right to a living personal and family wage. He maintained that women

who were forced by circumstance to support themselves and their children deserved equal compensation, but that, whenever possible, women should not work outside the home.

46. Ryan, "The Church and Birth Control," in *Family Limitation and The Church and Birth Control* (New York: Paulist Press, 1916), 16–19.
47. Reilly, *Surgical Solution*, 71–72; Harry H. Laughlin, *The Legal Status of Eugenical Sterilization* (Chicago: Psychopathic Laboratory of the Municipal Court of Chicago, 1929), 57.
48. Elizabeth K. McKeown, "Catholic Identity in America," in *America in Theological Perspective*, ed. Thomas M. McFadden (New York: Seabury Press, 1976), 59, 63–65. For more on the National Catholic War Council, see Elizabeth K. McKeown, *War and Welfare: American Catholics and World War I* (New York: Garland, 1988); and Douglas J. Slawson, *The Foundation and First Decade of the National Catholic Welfare Council* (Washington, DC: Catholic University of America Press, 1992), 26–35.
49. Ryan's biographer characterized the piece as "perhaps the most forward-looking social document ever to have come from an official Catholic agency in the United States" (see Francis L. Broderick, *Right Reverend New Dealer: John A. Ryan* [New York: Macmillan, 1963], 105). See also Joseph M. McShane, S.J., *"Sufficiently Radical": Catholicism, Progressivism, and the Bishops' Program of 1919* (Washington, DC: Catholic University of America Press, 1986).
50. Committee on Special War Activities, *Social Reconstruction: A General Review of Problems and a Survey of Remedies* (Washington, DC: National Catholic War Council, 1919), 24.
51. James Cardinal Gibbons, "The Pastoral Letter of 1919," in *The National Pastorals of the American Hierarchy, 1792–1919*, ed. Peter Guilday (Washington DC: National Catholic Welfare Council, 1923), 265–340.
52. Ibid., 307.
53. Ibid., 313.

NOTES TO CHAPTER 2

1. Barry Mehler, "A History of the American Eugenics Society, 1921–1940" (PhD diss., University of Illinois at Urbana–Champaign, 1988), 36–59.
2. Ibid., 58, 61.
3. Eugenics Committee Minutes, 9 August 1922, American Eugenics Society Papers, American Philosophical Society Library, Philadelphia [hereafter cited as AES Papers]; "The Advisory Council of the Eugenics Committee of the United States of America," *Eugenical News* 8, no. 4 (April 1923).
4. Mehler, "History of the American Eugenics Society," 68.
5. Ibid., 110.
6. "Report of the Sub-committee on the Ultimate Program to be Developed by the Eugenics Society of the United States of America," 73.
7. Ibid., 74.
8. Ibid., 78.
9. For a treatment of the popular and educational efforts of the eugenics movement, see Selden, *Inheriting Shame*, 39–83.
10. "Report of Sub-Committee on the Ultimate Program to be Developed by the Eugenics Society of the United States of America," 78–80.
11. Nancy Leys Stepan and Sander L. Gilman, "Appropriating the Idioms of Science: The Rejection of Scientific Racism," in *The Bounds of Race: Perspectives on Hegemony*

and Resistance, ed. Dominick LaCapra (Ithaca, NY: Cornell University Press, 1991), 73–103 (quotation, 94).

12. Eugenics Committee Minutes, 26 October 1923, AES Papers. The most comprehensive treatment of the CCC is contained in Christine Rosen, *Preaching Eugenics: Religious Leaders and the American Eugenics Movement* (Oxford: Oxford University Press, 2004), 114–20.
13. Rosen, *Preaching Eugenics*, 118–19.
14. US Department of Commerce and Labor, Bureau of the Census, *Census of Religious Bodies: 1926* (Washington, DC: US Government Printing Office, 1929), 2:1255–57; 1920 and 1930 state level census data, *Historical Census Browser*, University of Virginia Geospatial and Statistical Data Center, 2004, http://mapserver.lib.virginia.edu/.
15. This geographic convergence is significant given Elazar Barkan's comments in his discussion of scientific racism: "The core of the discipline [the study of race] was determined largely by geographical proximity to major academic centers. Distance and lack of communication continuously frustrated close interaction. Geography in and of itself was important; New York was the center of racial discourse in the United States. California was too far away from the center for even the most prominent of its scientists to participate in the national scientific-intellectual discourse on race" (Barkan, *Retreat of Scientific Racism*, 9).
16. A most helpful discussion of the chilling effect in the sciences of the condemnation of modernism is Appleby, "Exposing Darwin's 'Hidden Agenda,'" 173–208.
17. Elizabeth McKeown, "From *Pascendi* to *Primitive Man*: The Apologetics and Anthropology of John Montgomery Cooper," *US Catholic Historian* 13, no. 2 (Winter 1995).
18. John Montgomery Cooper, *Religion Outlines for Colleges*, vols. 1–4 (Washington, DC: Catholic Education Press, 1924–30).
19. Cooper, *Religion Outlines for Colleges*, vol. 1, *Course I. The Catholic Ideal of Life*, 2nd ed., rev. (Washington, DC: Catholic Education Press, 1937), 154.
20. Cooper, *Religion Outlines for Colleges*, vol. 4, *Course IV. Life Problems* (Washington, DC: Catholic Education Press, 1928), 61–93 (quotation, 68).
21. Broderick, *Right Reverend New Dealer*, 115.
22. "Report of Sub-committee on the Ultimate Program to be Developed by the Eugenics Society of the United States of America," 79.
23. "Prizes for Sermons on Eugenics," *Eugenical News* 11, no. 3 (March 1926); "Sermons on Eugenics," *Eugenical News* 11, no. 4 (April 1926). See also Rosen, *Preaching Eugenics*, 120–27.
24. Minutes of the Board of Directors, 16 November 1929, AES Papers.
25. Dolan, *The American Catholic Experience*, 245, 392.
26. [Charles B. Davenport,] "White America," *Eugenical News* 9, no. 1 (January 1924). A copy of the review is also located in the Charles B. Davenport Papers at the American Philosophical Society, Philadelphia.
27. "Fifteenth Meeting of the Eugenics Committee of the United States of America," 31 October 1925, Minutes (1925–1956), folder 1, 1925–1935, AES Papers.
28. "A Free Hand in the Philippines," *Eugenical News* 15, no. 5 (May 1930).
29. Dealing with tremendous diversity due to the immigration of the nineteenth century, Catholic bishops allowed the establishment of national parishes that recognized the authenticity of distinct and traditional styles of worship and that worked to preserve the ethnicity of immigrants from Southern and Eastern Europe. Instead

of forcing new immigrants to adopt the worship styles and culture of the predominantly Irish and German hierarchy of the church, the establishment of "national" parishes allowed new immigrants to participate in the life of the church without compromising their ethnic heritage. This is not to say that there were not serious instances of conflict between ethnic groups in Catholic urban centers, but the national parishes theoretically accommodated difference more than they discouraged it. This accommodation of difference was possible only because the individuals who populated those congregations shared a common commitment to the core elements of the Catholic faith and were united by a sacramental theology that centered on the Eucharist. Similarly, areas with historically Catholic populations, such as Louisiana, adhered to a system of racial differentiation that recognized a greater variation than the one-drop rule that dominated much of the United States by the 1920s. The key texts on these complicated issues are: R. Emmett Curran, "Splendid Poverty: Jesuit Slave-Holding in Maryland, 1805–1838," in *Catholics in the Old South: Essays in Church and Culture*, ed. Randall Miller and Jon Waklyn (Macon, GA: Mercer University Press, 1983), 125–46; Cyprian Davis, *The History of Black Catholics in the United States* (New York: Crossroads, 1990); McGreevy, *Parish Boundaries*; and David W. Southern, *John LaFarge and the Limits of Catholic Interracialism, 1911–1963* (Baton Rouge: Louisiana State University Press, 1996). On the one-drop rule, see Joel Williamson, *New People: Miscegenation and Mulattoes in the United States* (Baton Rouge: Louisiana State University Press, 1995), 20–24; and F. James Davis, *Who Is Black? One Nation's Definition* (University Park: Penn State University Press, 1991), 31–50.

30. Higham, *Strangers in the Land*, 300–324.
31. Reilly, *Surgical Solution*, 62–65; and Kevles, *In the Name of Eugenics*, 102–3. See also Kenneth Ludmerer, "Genetics, Eugenics, and the Immigration Restriction of 1924," *Bulletin of the History of Medicine* 46, no. 1 (January 1972); and Kenneth M. Ludmerer, *Genetics and American Society: A Historical Appraisal* (Baltimore: Johns Hopkins University Press, 1972), 100–13.
32. Ives to Ryan, 13 May 1929; Ryan to Ives, 18 May 1929; Ives to Ryan, 20 May 1929; and Ryan to Ives, 23 May 1929, John A. Ryan Papers, Archives of the Catholic University of America, Washington, DC [hereafter cited as Ryan Papers].
33. "Personal Factors in Immigration: Should They Get More Attention?," *Eugenics* 2, no. 7 (July 1929).
34. "The New Immigration Law," *Catholic Charities Review* 1 (April 1917); reprinted in the *Fortnightly Review*, 1 May 1916, 133–35. Ryan also expressed similar opinions in a paper he contributed to the Social Services Commission of the American Federation of Catholic Societies (quoted in John C. Murphy, *An Analysis of the Attitudes of American Catholics toward the Immigrant and the Negro, 1825–1925* [Washington, DC: Catholic University of American, 1940], 101).
35. Ives to Ryan, 9 December 1929; and Ryan to Ives, 17 December 1929, Ryan .
36. "Is Eugenics Racial Snobbery?" *Eugenics* 2, no. 2 (February, 1929).
37. Whitney to Cooper, 8 September 1930, John Montgomery Cooper Papers, Archives of the Catholic University of America, Washington, DC [hereafter cited as Cooper Papers].
38. Cooper to Whitney, 19 September 1930, Cooper Papers.
39. Hebert Spencer Jennings, "Undesirable Aliens," *The Survey* 51 (1923): 309–12, 364; and Raymond Pearl, "The Biology of Superiority," *American Mercury* 12 (November 1927): 260. See also Degler, *In Search of Human Nature*, 150–51; and Kevles, *In the Name of Eugenics*, 129–75.

40. Edwin Grant Conklin, "Some Recent Criticisms of Eugenics," *Eugenical News* 13, no. 5 (May 1928), 65.
41. "The New Conquest of California," *Eugenical News* 13, no. 2 (February 1928): 31.
42. For more on "race suicide," see chap. 1.
43. For more on "momism," see Paula J. Caplan, "Mother-Blaming," in *"Bad" Mothers: The Politics of Blame in Twentieth-Century America*, ed. Molly Ladd-Taylor and Lauri Umansky (New York: New York University Press, 1998); and Jennifer Terry, "'Momism' and the Making of Treasonous Homosexuals," ibid.
44. "The New Conquest of California."
45. "Feminism and the 'New Woman': Their Eugenical Significances," *Eugenics* 3, no. 8 (August 1930).
46. Ibid.
47. On feminism and social motherhood, see Peter Filene, *Him/Her/Self: Gender Identities in Modern America*, 3rd ed. (Baltimore: Johns Hopkins University Press, 1998), 42–59. For a specific treatment of pronatalism and eugenics, see Elaine Tyler May, *Barren in the Promised Land: Childless Americans and the Pursuit of Happiness* (Cambridge, MA: Harvard University Press, 1997), 61–95.
48. "Feminism and the 'New Woman.'"
49. Whitney to Cooper, 23 August 1929; and Cooper to Whitney, 27 August 1929, Cooper Papers.
50. Gerrard, "Eugenics and Catholic Teaching"; Ryan, "The Church and Birth Control"; and Somerville, "Eugenics and the Feeble-Minded."
51. Margaret Sanger, *The Pivot of Civilization* (1922; repr., London: J. Cape, 1923), 40.
52. Daniel Kevles aptly explains this transition: "Stripped of its assertive feminism, contraception became acceptable to conservative eugenicists, for there was natural harmony between their social predilections and the pro-birth-control rational advanced by Havelock Ellis: 'The superficially sympathetic man flings a coin to the beggar; the more deeply sympathetic man builds an almshouse for him so that he need no longer beg; but perhaps the most radically sympathetic of all is the man who arranges that the beggar shall not be born'" (Kevles, *In the Name of Eugenics*, 90).
53. The ongoing controversy between Catholics and birth controllers from 1914 through the 1960s has been well documented. The key text on this is Tentler's excellent *Catholics and Contraception*. See also Chesler, *Woman of Valor*; and Kathleen Tobin-Schlesinger, "Population and Power: The Religious Debate over Contraception, 1916–1936," 2 vols. (PhD diss., University of Chicago, 1994).
54. "The Reproduction Rate of Genius: Will Birth Control Diminish It?" *Eugenics* 2, no. 3 (March 1929).
55. Ives to Cooper, 27 March 1929, Cooper Papers. For the exchanges about birth control, see Whitney to Cooper, 7 March 1929; Cooper to Whitney, 12 March 1929; Whitney to Cooper, 10 June 1930; Cooper to Whitney, 12 June 1930; Whitney to Cooper, 16 June 1930; Popenoe to Cooper, 23 October 1923; Popenoe to Cooper, 25 July 1925; Whitney to Cooper, 16 January 1925; Cooper to Whitney, 20 January 1925; Whitney to Cooper, 23 March 1925; and Whitney to Cooper, 7 May 1925, ibid.
56. John Montgomery Cooper, *Birth Control* (Washington, DC: National Catholic Welfare Council, 1923); "News and Notes," *Eugenical News* 10, no. 3 (March 1925).
57. Cooper to Ives, 1 April 1929, Cooper Papers.
58. John Montgomery Cooper, "The Birth Rate of Genius: Does Contraception Curb It?," *Eugenics* 2, no. 5 (May 1929).

59. Edward M. East, "The Birth Rate of Genius: Does Contraception Curb It?" *Eugenics* 2,no. 5 (May 1929).
60. For the *American Ecclesiastical Review* exchange, see chap. 1.
61. John A. Ryan, *Human Sterilization* (Washington, D.C.: National Catholic Welfare Conference, 1927).
62. "Refusal, Seton Hall College," *Eugenics* 3, no. 4 (April 1930).
63. Whitney to Cooper, 27 March 1930, Cooper Papers.
64. Cooper to Whitney, 8 April 1930, Cooper Papers.
65. Roswell Johnson worked at the University of Pittsburgh and was Popenoe's coauthor for the most widely used eugenics textbook, *Applied Eugenics*.
66. "Sterilization: Its Legality, Need; A Catholic and Some Eugenicists Speak," *Eugenics* 3, no. 5 (May 1930).
67. Pius XI, *Casti Connubii* (Encyclical of Pope Pius XI on Christian Marriage . . . , 31 December 1930), par. 71, http://www.vatican.va/holy_father/pius_xi/encyclicals/documents/hf_p-xi_enc_31121930_casti-connubii_en.html.
68. Clipping of newspaper article, 18 April 1931; and Ryan to Whitney, 9 April 1931, Ryan Papers.
69. Whitney to Ryan, 17 April 1931, Ryan Papers.
70. Whitney to Ryan, 17 April 1931, Ryan Papers.
71. Cooper to Whitney, 9 April 1931, Cooper Papers.
72. Whitney to Cooper, 21 April 1931, Cooper Papers.

NOTES TO CHAPTER 3

1. Ryan, *Human Sterilization*, 5. A copy of the pamphlet is located in National Catholic Welfare Conference Papers, Archives of the Catholic University of America, Washington, DC [hereafter cited as NCWC Papers].
2. Laughlin, *The Legal Status of Eugenical Sterilization*, 7.
3. Reilly, *Surgical Solution*, 84; Laughlin, *The Legal Status of Eugenical Sterilization*, 57–59; and Paul A. Lombardo, *Three Generations, No Imbeciles: Eugenics, the Supreme Court, and "Buck v. Bell"* (Baltimore: Johns Hopkins University Press, 2008), 293–94. In 1927, the list of states with valid laws included Alabama, California, Colorado, Connecticut, Delaware, Idaho, Indiana, Iowa, Kansas, Maine, Michigan, Minnesota, Montana, Nebraska, Nevada, New Hampshire, North Carolina, North Dakota, Oregon, South Dakota, Utah, Virginia, Washington, and Wisconsin. The southern states were late to warm to the idea of eugenic sterilization legislation. Though laws were considered in Louisiana during the mid-1920s, other southern states debated the matter through the 1930s and well into the 1940s (see Larson, *Sex, Race, and Science*).
4. Noll, *Feeble-Minded in Our Midst*, 1–4.
5. Stephen Jay Gould has effectively explored and debunked Goddard's data and conclusions from the Kallikak study, noting that by the late 1920s Goddard had recanted his stance on feeble-mindedness being incurable, uneducable, or necessitating institutionalization (see Gould, *The Mismeasure of Man*, 158–74). For more on family studies, see Nicole Hahn Rafter, ed., *White Trash: The Eugenic Family, 1877–1919* (Boston: Northeastern University Press, 1986).
6. Harry H. Laughlin, *Eugenical Sterilization in the United States* (Chicago: Psychopathic Laboratory of the Municipal Court of Chicago, 1922); and Laughlin, *Eugenical Sterilization: 1926: Historical, Legal, and Statistical Review of Eugenical Sterilization in the United States* (New Haven, CT: American Eugenics Society, 1926), 2–3.

7. Gibbons, "Pastoral Letter of 1919," 296–97. On the establishment of the National Catholic Welfare Conference, see Slawson, *Foundation and First Decade of the National Catholic Welfare Council*, 45–83.
8. Douglas J. Slawson, *The Department of Education Battle, 1918–1932: Public Schools, Catholic Schools, and the Social Order* (Notre Dame, IN: University of Notre Dame Press, 2005), 5.
9. "Bishops Bring United Counsel to Problems of National Importance," NCWC Bulletin 2 (October 1920): 1.
10. Slawson, *Foundation and First Decade of the National Catholic Welfare Council*, 70–83.
11. The legislative activity in New Jersey in 1926 reflected a resurgence of interest in the eugenic benefits of sterilization. In 1911, the state legislature had passed and Woodrow Wilson (then governor and future president) had signed a statute that provided for the sterilization of inmates of state reformatories and charitable or penal institutions. The law targeted persons who, in the evaluation of a state board of examiners, had conditions that were unlikely to improve, thus making their reproduction inadvisable. Though there were five state institutions that qualified for sterilizing inmates under the law, the statute faced a constitutional challenge before any operations were done. On November 18, 1913, the New Jersey State Supreme Court declared the law unconstitutional on the grounds that it violated the equal protection clause of the Fourteenth Amendment. By the mid-1920s, the eugenic sterilization lobby had regained its momentum and was again encouraging the state legislature to pass a sterilization statute (see Laughlin, *Eugenical Sterilization in the United States*, 10, 80, 174).
12. Founded in 1920, the National Council of Catholic Women was an outgrowth of the National Catholic War Council's Committee on Women's Activities. Under the auspices of the Welfare Conference's Department of Lay Organizations, directed by Cleveland Bishop Joseph Schrembs, the NCCW coordinated and connected the various laywomen's societies and organizations throughout the country. The NCCW sustained a significant and vibrant connection among women's groups around the country. During the decades between its founding and World War II, the council took an active role in many issues of the day. It opposed the persecution of the church in Mexico; fought against the Sheppard-Towner Act; took a stand against the distribution of contraception and contraceptive information; participated in immigrant Americanization and education work; established community houses; and pursued social justice causes and family issues. In addition to working on these issues, beginning in 1921, the NCCW sponsored the National Catholic School of Social Service, which trained a legion of Catholic social workers. Such activities made the women of the NCCW integral participants in the Catholic effort to thwart eugenic sterilization statutes (see Ruth O'Halloran, "Organized Catholic Laywomen: The National Council of Catholic Women, 1920–1995" [PhD diss., Catholic University of America, 1995], chap. 1–3).
13. The information on the ratio of Catholics to the total population of the states was calculated using 1920 and 1930 state level census data from the *Historical Census Browser* at the University of Virginia Geospatial and Statistical Data Center; and from US Department of Commerce and Labor, Bureau of the Census, *Census of Religious Bodies: 1926*, 2:1256. The New Jersey total population in 1926 was approximately 3.76 million persons, and the Catholic population that year was 1.06 million persons.
14. Mary G. Hawks to John J. Burke, 6 February 1926, Collection 121, NCWC Papers.

15. Burke to Hawks, 12 February 1926, Collection 121, NCWC Papers.
16. For more on Catholic challenges to eugenics in Vermont, see Gallagher, *Breeding Better Vermonters*, 78–79, 84.
17. Burke to Rev. William P. Crosby, 2 February 1927, Collection 121D, NCWC Papers.
18. NCWC News Service, "State Legislatures Discuss Anti-Evolution Measures, Propose Marriage Reforms," *The Tidings*, 25 February 1927, 6; NCWC News Service, "State Legislatures Discuss Bills of Interest to Catholics," *The Tidings*, 4 March 1927, 7; NCWC News Service, "Quantity Production Marks Output in Legislatures of Many States," *The Tidings*, 25 March 1927, 4; NCWC News Service, "State Legislatures Struggle with Several Eugenic Measures," *The Tidings*, 22 April 1927, 13.
19. The best account of nativism in the United States continues to be Higham, *Strangers in the Land*. The tendency for non-Catholics to question the fitness of Catholics to participate in liberal democracy is long-standing, and deftly examined by John T. McGreevy in *Catholicism and American Freedom: A History* (New York: W. W. Norton, 2003).
20. Slawson, *Department of Education Battle*, 35–76.
21. Smith retired at the end of the 66th Congress, and Senator Sterling replaced him as a sponsor of a slightly different education bill during the following session.
22. John A. Ryan, *The State and the Church* (Washington, DC: National Catholic Welfare Conference, 1922), 42–49.
23. On education, see, e.g., the battle over school laws in the state of Oregon (Thomas J. Shelly, "The Oregon School Case and the National Catholic Welfare Conference," *Catholic Historical Review* 75, no. 3 [1989]). On the controversy over contraception, see Chesler, *Woman of Valor*; Tobin-Schlesinger, "Population and Power"; Leslie Woodcock Tentler, "'The Abominable Crime of Onan': Catholic Pastoral Practice and Family Limitation, 1875–1919," *Church History* 71, no. 2 (June 2002); and Tentler, *Catholics and Contraception*.
24. Ryan, *The State and the Church*, 207.
25. Ibid., 276–77.
26. National Council of Catholic Men, *Special Bulletin on Ohio Sterilization and Marriage License Bills* (Washington, D.C.: National Catholic Welfare Conference, 1927), 1–2, in Collection 121D, NCWC Papers.
27. Charles F. Dolle to Bishop Joseph Schrembs, 31 March 1927, Joseph Schrembs Papers, Archives of the Diocese of Cleveland, Cleveland, OH [hereafter cited as Schrembs Papers].
28. National Council of Catholic Men, *Special Bulletin on Ohio Sterilization and Marriage License Bills*, 1–2, in Collection 121D, NCWC Papers.
29. Bishop Schrembs's correspondence with Dolle about the Ohio legislation, and the subsequent production of the "Information Bulletin," provided a significant link between the national and the local structures of US Catholicism. Bishop Schrembs was chairman of the Department of Lay Organizations, which oversaw the National Council of Catholic Men and monitored Dolle's administration of the day-to-day affairs of the organization. Through the Department of Lay Organization, Schrembs had guided the foundation and development of both the NCCM and the NCCW to foster lay activism in local dioceses around the country. While the NCCW developed a vibrant national organization, the NCCM functioned more directly under Schrembs's control. The bishop envisioned the organization as being deeply involved with social issues, such as the fight against birth control, the campaign for parochial schools, the care of immigrants, and the education of Catholic men on

social and moral teaching. The sterilization law in Ohio allowed him to exercise his authority as spokesperson for the Catholics within the Diocese of Cleveland and to oversee the involvement of the NCCM in the hopes of providing materials to educate the laity about the variety of reasons for opposing eugenics policy initiatives (see Martin Frank Poluse, "Archbishop Joseph Schrembs and the Twentieth Century Catholic Church in Cleveland, 1921–1945" [PhD diss., Kent State University, 1991], 250–55).

30. The bulletin drew from the work of Dr. William E. Fernald, former superintendent of the Massachusetts School for Feebleminded, and Dr. George K. Pratt, Assistant to the Medical Director of the National Committee for Mental Health. Fernald published the first medical critique of eugenics in 1919, based on following the lives of 646 non-sterilized feeble-minded persons who had been discharged from a state institution in Massachusetts. They exhibited a low marriage rate and a low birth rate, contrary to the claims of sterilization advocates. Fernald argued that eugenicists mistook a high birthrate among the poor and uneducated for that of the mentally challenged (see Reilly, *Surgical Solution*, 122).
31. National Council of Catholic Men, *Special Bulletin on Ohio Sterilization and Marriage License Bills*, 9, in Collection 121D, NCWC Papers.
32. Leo XIII, *Rerum Novarum* (Encyclical of Pope Leo XIII on Capital and Labor, 15 May 1891), par. 12–14, http://www.vatican.va/holy_father/leo_xiii/encyclicals/documents/hf_l-xiii_enc_15051891_rerum-novarum_en.html.
33. Handwritten note on Dolle to Schrembs, 31 March 1927; McFadden to Ryan (telegram sent to members of the Health Committee), n.d.; and telegram to John A. Hadden, State House, 8 April 1927, Schrembs Papers.
34. McNicholas's statement to Ohio State Legislature, quoted in NCWC News Service, "State Legislatures Struggle with Several Eugenic Measures," 13.
35. Quoted in "Sterilization Bill is Killed," *Catholic Universe Bulletin*, 15 April 1927.
36. "Sterilization Bill is Killed," 1.
37. Kevles, *In the Name of Eugenics*, 113–47; and Reilly, *Surgical Solution*, 111–27.
38. The most comprehensive work on the *Buck v. Bell* case has been done by Paul Lombardo in *Three Generations, No Imbeciles* (see esp. 103–56). My narrative of these events is drawn from Lombardo's work, and from J. David Smith and K. Ray Nelson, *The Sterilization of Carrie Buck* (Far Hills, NJ: New Horizon, 1989).
39. Quoted in Smith and Nelson, *Sterilization of Carrie Buck*, 32–33.
40. During Gordon's period of deliberation, Priddy died of Hodgkin's disease. His assistant, Dr. J. H. Bell, assumed his position as head of Virginia's State Colony for Epileptics and Feeble-Minded.
41. Buck v. Bell, 274 U.S., 47 Sup. Ct. 584, 585 (1927).
42. Patrick J. Ward, "Sterilization of Mental Defectives in the State of Virginia" (report), 6 May 1927; and Ward to Burke, memo regarding sterilization in Virginia, 7 May 1927, Collection 121D, NCWC Papers.
43. The relative numbers of Catholics in Virginia in 1926 was calculated using 1920 and 1930 state level census data from the *Historical Census Browser*, University of Virginia Geospatial and Statistical Data Center; and from US Department of Commerce and Labor, Bureau of the Census, *Census of Religious Bodies: 1926*, 2:1256. For a summary of the religious history of the Commonweal of Virginia, see W. Harrison Daniel, "Virginia," in *Religion in the Southern States*, ed. Samuel S. Hill (Macon, GA: Mercer University Press, 1983).
44. Lombardo, *Three Generations, No Imbeciles*, 174–77.

45. "A New Issue," *Catholic Universe Bulletin*, 6 May1927, 16.
46. Philip Reilly notes that within a few years of the *Buck* decision, the number of states with sterilization statutes increased from seventeen to thirty. Additionally, while the total number of institutionalized persons who had been sterilized between 1907 and 1927 was 8,515, that number had jumped to a total of 38,087 persons by 1941 (see Reilly, *Surgical Solution*, 87, 97).
47. "Unjustified Sterilization."
48. Francis S. Betten, S.J., "Sterilization Banned by Natural Law, Says Jesuit," *Catholic Universe Bulletin*, 20 May 1927. Betten's image of a female victim of compulsory sterilization pointed to the central place of motherhood in Catholic social teaching, which is also prominent in the discussions of wage justice in *Rerum Novarum*.
49. Reilly, *Surgical Solution*, 97–98.
50. Dolle to Burke, 28 May 1927; and Burke memo, 3 June 1927, Collection 121D, NCWC Papers.
51. Dolle to Burke, 8 June 1927; and "Petition for Rehearing and Argument" for *Buck v. Bell*, Collection 121D and 121, respectively, NCWC Papers.
52. The questions that accompanied Al Smith's campaign for the presidency, about whether or not a Catholic was fit to hold executive office, prompted John Ryan to draft another text explaining the Catholic teachings on church and state (see John A. Ryan, *The Catholic Church and the Citizen* [New York: Macmillan, 1928]).
53. "Brush Foundation Deems Humans Barnyard Animals, Bishop Says," *Catholic Universe Bulletin*, 22 June 1928; "Eugenics: True and False," *Catholic Universe Bulletin*, 6 July 1928.
54. Ryan, *Human Sterilization*, 5.
55. Burke memo to self, 30 October 1928; and Burke to Ward, 24 January 1929, Collection 121, NCWC Papers.
56. Ezra S. Gosney and Paul Bowman Popenoe, eds., *Sterilization for Human Betterment: A Summary of Results of 6,000 Operations in California, 1909–1929* (New York: Macmillan, 1929).
57. Raymond A. McGowan to Burke, 18 February 1929, Collection 121D, NCWC Papers.
58. Ward to Burke in re: "Evil Results of Sterilization," 5 April 1929, Collection 121D, NCWC Papers.
59. Ibid.
60. Ulrich A. Hauber, *Inheritance of Mental Defect* (Washington, DC: National Catholic Welfare Conference, 1930).
61. Charles Bernstein, M.D., *Social Care of the Mentally Deficient* (Washington, DC: National Catholic Welfare Conference, 1930).
62. William F. Montavon, *Eugenic Sterilization in the Laws of the States* (Washington, DC: National Catholic Welfare Conference, 1930).
63. "Study Clubs Are Urged by D.C.C.W.," *Catholic Universe Bulletin*, 21 June 1929; "Why Study Clubs?" *Catholic Universe Bulletin*, 8 August 1930.
64. Pius XI, *Casti Connubii*, par. 71.
65. John A. Ryan, *Moral Aspects of Sterilization* (Washington, DC: National Catholic Welfare Conference, 1930), 6, 21–22.

NOTES TO CHAPTER 4

1. Paul Popenoe, "A Roman Catholic View of Sterilization," *Journal of Social Hygiene* 14, no. 3 (March 1928).

2. Ibid., 190–91.
3. Joseph Mayer, S.J., "Eugenics in Roman Catholic Literature," *Eugenics* 3, no. 2 (February 1930), 44.
4. Ibid., 46.
5. Ibid.
6. Ibid.
7. Pius XI, *Casti Connubii*, par. 4.
8. Noonan, *Contraception*, 463–70, 482–86.
9. Ibid., 486–504.
10. "Resolution 15: The Life and Witness of the Christian Community—Marriage and Sex," Lambeth Conference, Resolutions from 1930, http://www.lambethconference.org/resolutions/1930/1930-15.cfm.
11. Noonan, *Contraception*, 504–6.
12. Pius XI, *Casti Connubii*, par. 63–71.
13. Ibid., par. 67.
14. Noonan, *Contraception*, 512.
15. Pius XI, *Casti Connubii*, par. 68.
16. Ibid., par. 69.
17. Ibid., par. 70.
18. E.g., see "Modern Marital Evils," *Catholic Universe Bulletin*, 16 January 1931).
19. "Encyclical of Pope to be Broadcast," *The Tidings*, 30 January 1931, 1.
20. "Copy of Encyclical for Each of 1,000 Families in Parish," *The Tidings*, 30 January 1931, 3.
21. John A. Ryan, "Moral Teaching of the Encyclical on Christian Marriage," *American Ecclesiastical Review* 84, no. 9 (March 1931), 268.
22. Hudson Hawley, "Pope Pius Will Issue Encyclical against Birth Control Spread," *Atlanta Constitution*, 5 January 1931; "Pope Strict with Cupid," *Los Angeles Times*, 5 January 1931; "Expect Pope to Tighten Rules on Marriage," *Chicago Daily Tribune*, 5 January 1931.
23. "Pontiff Proclaims Sanctity of Matrimony," *Los Angeles Times*, 9 January 1931.
24. "Local Minister Praises, Criticizes Teachings of Pope in Encyclical," *Atlanta Constitution*, 12 January 1931.
25. Edmond Taylor, "Pope Flays Sex Codes of Today as Deadly Sins," *Chicago Daily Tribune*, 9 January 1931.
26. "Rome Has Spoken!," *The Christian Century*, 4 February 1931.
27. Ibid.
28. "Pope Pius XI's Encyclical on Marriage," *Eugenical News* 16, no. 2 (February 1931).
29. Roswell Johnson, "Legislation," *Eugenics* 4, no. 2 (February 1931); Kenneth C. MacArthur, "Eugenics and the Church," *Eugenics* 4, no. 2 (February 1931).
30. See Mark B. Adams, ed., *The Wellborn Science: Eugenics in Germany, France, Brazil, and Russia* (New York: Oxford University Press, 1990); Gunnar Broberg and Nils Roll-Hansen, eds., *Eugenics and the Welfare State: Sterilization Policy in Denmark, Sweden, Norway, and Finland* (East Lansing: Michigan State University Press, 1996); and Nancy Leys Stepan, *"The Hour of Eugenics": Race, Gender, and Nation in Latin America* (Ithaca, NY: Cornell University Press, 1991). Eugenics in the Scandinavian countries was deeply linked to the process of modernization and the growth of efficient welfare states. Eugenics served as a justification for a range of measures, from alcohol restrictions to sterilization, designed to protect the welfare of the race in Norway. Scientists in Denmark were heavily influenced by the work of Henry Goddard and

the belief in hereditary determinism. Marriage restrictions and eugenic sterilization were presented as a necessary part of a larger reform agenda. Like Denmark, the eugenicists in Sweden pursued state efficiency and preemptive social reform through sterilization. These plans encouraged the growth of a productive national population. Together, the history of eugenics in the Scandinavian countries provides a picture of the ways that a firm belief in strict hereditary determinism and Mendelian genetics could be mobilized in support of a host of measures to craft an efficient, racially unified national population.

In both France and Latin America, where the medical profession had more clout, and where Lamarck's theory of evolution, based on the inheritance of acquired characteristics rather than on random variation (as in Darwinian theory), softened the eugenics movement. In France, a broad coalition of eugenics supporters in the Lamarkian mode focused on environmental and social issues that might be detrimental to the population. Their emphasis on improving health, social reform, and pronatalism resulted in a mostly positive cast to the movement there.

The neo-Lamarckism of French eugenics was a significant influence on the movement in Latin America. Eugenics supporters in Brazil talked about Mendel and Galton, but their working understanding of heredity was neo-Lamarckian, which convinced them that a proactive public policy could significantly improve the population. Sanitation, antipoverty reforms, social hygiene, and temperance efforts were all considered eugenic efforts in Brazil. In other Latin American countries, such as Cuba, Mexico, and Argentina, the eugenics movement also exhibited neo-Lamarckian aspects, focusing on the control and elimination of "race poison." By the 1930s, a version of negative eugenics had developed in Latin America, but it employed marriage restrictions to control reproduction, rather than the elaborate sterilization programs that characterized eugenics movements in other nations.

31. Kühl, *The Nazi Connection: Eugenics, American Racism, and German National Socialism* (New York: Oxford University Press, 2002), 19–20.

32. Ibid., 23–26.

33. Ibid., 29.

34. There is a vast and contradictory literature about the history of the Roman Catholic Church and the Holocaust, much of it centered on the actions and diplomacy of Pius XII. See Randolph L. Braham, ed., *The Vatican and the Holocaust: The Catholic Church and the Jews during the Nazi Era* (New York: Columbia University Press, 2000); John Cornwell, *Hitler's Pope: the Secret History of Pius XII* (New York: Viking, 1999); Daniel Jonah Goldhagen, *A Moral Reckoning: The Role of the Catholic Church in the Holocaust and Its Unfulfilled Duty to Repair* (New York: Alfred A. Knopf, 2002); Ralph M. McInerny, *The Defamation of Pius XII* (South Bend, IN: St. Augustine's Press, 2001); Michael Phayer, *The Catholic Church and the Holocaust, 1930–1965* (Bloomington: Indiana University Press, 2000); José M. Sánchez, *Pius XII and the Holocaust: Understanding the Controversy* (Washington, DC: Catholic University of America Press, 2002); and Garry Wills, *Papal Sin: Structures of Deceit* (New York: Doubleday, 2000), 13–69.

35. The church's stance rejecting racism is also suggested by the history of what has come to be known as the "hidden encyclical." In 1938, Pope Pius XI asked the American interracial activist, John Lafarge, S.J., to draft a letter condemning the racial policies of the Nazis. Pius XI died the following year, and Lafarge's draft, *Humani Generis Unitas*, never left the Vatican. While the document repeats many of the positions that Lafarge promoted in his efforts to ease race relations for African

Americans in the United States, it also includes instances of religious anti-Semitism. For an exploration of the history of the encyclical and the reasons for its suppression, see George Passelecq and Bernard Suchecky, *The Hidden Encyclical of Pius XI*, trans. Steven Rendell (New York: Harcourt Brace, 1997).

36. Donald J. Dietrich, "Catholic Resistance to Biological and Racist Eugenics in the Third Reich," in *Germans against Nazism: Nonconformity, Opposition, and Resistance in the Third Reich*, ed. Francis R. Nicosia and Lawrence D. Stokes (New York: Berg, 1990).
37. Donald J. Dietrich, "Catholic Eugenics in Germany, 1920–1945: Hermann Muckermann, S.J. and Joseph Mayer," *Journal of Church and State* 34, no. 3 (Summer 1992), 587–91.
38. In addition to this press coverage of the developing situation in Germany, religious activists in the United States recognized that they might be able to count on Catholics to support their efforts to protect Jews in Poland. In January 1937, the Reverend John Howland Lathrop from the American Committee on Religious Rights and Minorities approached John Ryan about making an ecumenical visit to Poland. The plan was to send one Catholic, one Jew, and one Protestant. Lathrop explained the mission as follows:

    > One fifth of the Jews of the world live in Poland. Nazi influence and a certain weakness of the Polish government makes us fearful that very shortly there will being action against the Jew in Poland similar to that which was taken in Rumania and Germany. We believe that Poland is sensitive to American public opinion because of America's influential part in setting up her independence after the World War. We believe, therefore, that a visitation from distinguished Americans to inquire into the situation at this time may help to prevent the tragedy which we foresee. Since Poland is essentially a Roman Catholic country we are anxious that the Catholic representative on the commission should be a person of the distinction and influence of yourself. (Lathrop to Ryan, 2 January 1937, Ryan Papers)

    Ryan declined the invitation, claiming that he could not possibly get a leave of absence from his duties at the Catholic University of America: "I greatly regret that this is the situation, as I should like to make some contribution toward the solution of the Polish Jewish problem. But it is impossible" (Ryan to Lathrop, 5 January 1937, Ryan Papers).
39. Quoted in Patrick J. Ward, "The Grave Issue of Sterilization," *Catholic Action*, February 1934, 31.
40. "Human Rights in Germany," *America*, 6 January 1934.
41. Allan Chase, *The Legacy of Malthus: The Social Costs of the New Scientific Racism* (Urbana: University of Illinois Press, 1980), 350.
42. Ward, "The Grave Issue of Sterilization."
43. James J. Walsh, "Race Betterment," *Commonweal*, 2 February 1934.
44. "The Pilgrim," *America*, 6 July 1935.
45. "Our Common Catholic Interests," *Catholic Action*, July 1935.
46. "Sterilization Law in Germany," *American Ecclesiastical Review* 35, no. 1 (July 1936) (quotations, 52, 64).
47. "Sterilization Conflicts," *Commonweal*, 12 April 1935.
48. "Attitude of Church on Sterilization Reaffirmed by Puerto Rican Prelate," *Catholic Action*, May 1934.
49. "Bishop Byrne Denounces 'Eugenic' Laws Adopted by Puerto Rican Assembly," *Cath-*

olic Action, September 1937; "Sterilization Condemned as Contrary to the Natural Law," *Catholic Action*, September 1937.

50. Laura Briggs, *Reproducing Empire: Race, Sex, Science, and U.S. Imperialism in Puerto Rico* (Berkeley: University of California Press, 2002), 81–89.
51. Barrett McGurn, "Three Vicious Laws Effective in Puerto Rico: An 'Easy' Way of Killing an Excess Population," *America*, 24 July 1937 (Ramos quote, 365).
52. Ibid.
53. Ibid., 364–65.

NOTES TO CHAPTER 5

1. Marion S. Norton, "An Account of the Work for Sterilization in New Jersey, 1934–1940, and a Presentation of the Need for a National Organization," 13 March 1941, box 1, folder 4 ("SLNJ, Corresp. [1940–1943]—Roman Catholic Opposition"), Association for Voluntary Sterilization Papers, Social Welfare History Archives, University of Minnesota [hereafter cited as AVSP Papers]. The author used the last name "Norton" early in her sterilization advocacy, prior to her second marriage, after which she adopted the name "Olden." For consistency's sake, I have referred to her as Olden within the text.
2. Though he did not contact Cooper directly concerning the bill, Schrembs drew on a letter that Cooper had written on Schrembs's request in 1929. In February of that year, Schrembs had approached Cooper for material to use in response to a newspaper article about the problems associated with feeble-mindedness and crime. Cooper responded with lengthy memorandum addressing the points raised in the article. Although Schrembs ultimately decided not to respond to the newspaper article, he eventually used Cooper's memo as the basis for his letter to the legislators (John Montgomery Cooper to Joseph Schrembs, 4 February 1929; and Schrembs to Cooper, 13 February 1929, Schrembs Papers).
3. Schrembs, form letter sent to all Ohio senators and representatives, 26 January 1931, Schrembs Papers.
4. Ortha O. Barr, Lima, to Schrembs, 3 February 1931, Schrembs Papers.
5. Nils P. Johnson, Youngstown, to Schrembs, 9 February 1931, Schrembs Papers.
6. Robert T. Secrest to Schrembs, 10 February 1931, Schrembs Papers.
7. Lawrence A Welsh, Steubenville, to Schrembs, 30 January 1931; A. S. Harding, Cleveland, to Schrembs, 30 January 1931; and Joseph N. Ackemen, Cleveland, to Schrembs, 30 January 1931, Schrembs Papers.
8. James C. Foster, Cleveland, to Monsignor James A. McFadden, 31 January 1931, Schrembs Papers. Population figures were derived from US Department of Commerce and Labor, Bureau of the Census, *Census of Religious Bodies: 1926*, 2:1256; and from US Department of Commerce and Labor, Bureau of the Census, *Census of Religious Bodies: 1936*, (Washington DC: Government Printing Office, 1941), 1:1531.
9. Sara A. Varley to Schrembs, 12 February 1931, Schrembs Papers.
10. Brown and McKeown, *The Poor Belong to Us*, 60–61, 152–63. On Catholics and the New Deal, see also Kenneth J. Heineman, *A Catholic New Deal: Religion and Reform in Depression Pittsburgh* (University Park: Pennsylvania State University Press, 1999); and David J. O'Brien, *American Catholics and Social Reform: The New Deal Years* (New York: Oxford University Press, 1968), 47–69.
11. Schrembs to the *Youngstown Vindicator* and the *Youngstown Telegram*, 13 February 1931, Schrembs Papers.
12. Schrembs to Bishop James J. Hartley, 17 February 1931, Schrembs Papers.

13. Schrembs to Senators L. L. Marshall, Joseph N. Ackeman, Laurence H. Norton, W. A. Greenlund, and James A. Reynolds, 10 March 1931, Schrembs Papers.
14. Lillian M. Westropp and Alice Kelly to the Hon. Marvin C. Harrison and the senators of Cuyahoga, Akron, and Youngstown, 25 January 1933, Schrembs Papers.
15. Shrembs to Senators Waldvogel, McManus, Matthews, Annat, LeFever, Lowery, and Pfieffer, 30 January 1933, Schrembs Papers.
16. Alice F. Kelly to Schrembs, 29 January 1933, Schrembs Papers.
17. Schrembs to W. H. Annat, 2 February 1933, Schrembs Papers.
18. Reilly, *Surgical Solution*, 49, 97.
19. Ibid., 94–102 (yearly totals are from the chart on p. 97).
20. "The Sterilization Craze," *America*, 14 March 1931.
21. Chris J. McSweeney, *The Case against Sterilization of Mental Defectives* (St. Louis: Central Bureau, Catholic Central Verein of America, 1931).
22. "Combating Sterilization Bills," *Central-Blatt and Social Justice* 24 (April 1931).
23. "Bettering the Stock," *America*, 6 October 1934.
24. "Sterilization in the United States," *Commonweal*, 20 September 1935.
25. "Salutary Words," *Commonweal*, 15 March 1935.
26. Paul Blakely, S.J., "Sterilization and Other Conjectures," *America*, 16 March 1935.
27. Paul Blakely, S.J., "Nams, Kallikaks, and Jukeses," *America*, 28 March 1936. The report from the American Neurological Association committee was eventually published in book form, under the title *Eugenical Sterilization: A Reorientation of the Problem* (New York: Macmillan, 1936).
28. NCWC News Service, "Inquiry Is Demanded into Sterilization of Girls' School Inmates," 1 November 1937, Social Action Department, NCWC Papers.
29. Jacob H. Landman, *Human Sterilization* (New York: Macmillan, 1932).
30. Ryan to J. H. Landman, 27 May 1932, Ryan Papers.
31. Jacob H. Landman, "Race Betterment by Human Sterilization," *Scientific American*, June 1934.
32. Ezra S. Gosney, "Eugenic Sterilization: Human Betterment Demands It," *Scientific American*, July 1934.
33. C. Thomalla, M.D., "The Sterilization Law in Germany," *Scientific American*, September 1934.
34. Ignatius W. Cox, S.J., "The Folly of Human Sterilization," *Scientific American*, October 1934.
35. Kevles, *In the Name of Eugenics*, 122–28; and Barkan, *Retreat of Scientific Racism*, 191–220.
36. Paul Popenoe, "A Debate on Sterilization: Intelligent Eugenics," *The Forum*, July 1935.
37. Henry Goddard Leach to Ryan, 23 April 1935, Ryan Papers.
38. Frank A. Hall to John J. Burke, 14 August 1934, NCWC Papers.
39. Larson, *Sex, Race, and Science*, 119–45.
40. "Table 3.—Number and Membership of Churches . . . by States, 1936," in US Department of Commerce and Labor, Bureau of the Census, *Census of Religious Bodies: 1936*, 2:1531; and 1930 and 1940 state level census data, *Historical Census Browser*.
41. William L. Truby, "House Passes Sterilization Bill," *Decatur Daily*, 10 May 1935, 1.
42. The historian Edward J. Larson has written eloquently about this legislative battle in his account of eugenics in the southern United States, and I am indebted to him for uncovering the sources surrounding these events (see Larson, *Sex, Race, and Science*, 139–45).
43. "Sterilization in Alabama," *Montgomery Advertiser*, 6 February 1935, 4.

44. "Father Obering Attacks Alabama Sterilization Law in Birmingham," *The Spring-Hillian*, 29 April 1935, 1.
45. William F. Obering, S.J., "Authority Says Alabama Sterilization Measure Is Declaration of Slavery," *Catholic Week* (Birmingham), 5 May 1935, 1, 6, 7.
46. Truby, "House Passes Sterilization Bill," 1; Ralph Hurst, "$2000 Tax-Free Homesteads Is Urged in House," *Birmingham News*, 10 May 1935, 1; "The House Passes the Sterilization Bill," *Birmingham News*, 11 May 1935, 4; "Sterilization Progresses," *Montgomery Advertiser*, 11 May 1935, 4; and Frank Gordy, "House Spotlight Is Taken by Sterilization Measure," *Mobile Register*, 11 May 1935, 1.
47. "Solons Hear Opposition to Sterilization," *Catholic Week* (Birmingham), 19 May 1935, 1, 2; and Larson, *Sex, Race, and Science*, 142.
48. Population comparisons are derived from the data for Alabama in volume 2 of US Department of Commerce and Labor, Bureau of the Census, *Census of Religious Bodies: 1936*, which lists the Roman Catholic population as 33,104 (p. 1533), the Baptist population as 212,855 (p. 118), and the Methodist Episcopal Church, South, as 155,416 (p. 1138).
49. Hugh Sparrow, "Sterilization Bill for State Up to Governor," *Birmingham Age–Herald*, 1 June 1935, 1; "Sterilization Bill Passes Senate," *Montgomery Advertiser*, 1 June 1935, 1, 3; "Sterilization Bill Passes, Awaits Governor's Action," *Mobile Register*, 1 June 1935, 1; and Larson, *Sex, Race, and Science*, 141.
50. "Sterilization Bill Killed by Veto," *Montgomery Advertiser*, 26 June 1935, 1 (Graves quote); "Senate Approves Decision to Boost Income Tax," *Decatur Daily*, 26 June 1935, 1; and "Sterilization Bill Gets Veto of Governor Graves," *Mobile Register*, 26 June 1935, 1.
51. The article in the July 4, 1935, edition of the *Alabama Baptist* is quoted in Wayne Flynt, *Alabama Baptists: Southern Baptists in the Heart of Dixie* (Tuscaloosa: University of Alabama Press, 1998), 391.
52. Graves's statement is quoted in J. N. Baker, "Medical and Health Legislation in 1935," *Journal of the Medical Association of the State of Alabama* 5 (October 1935).
53. William L. Truby, "Sedition Bill Veto Is Under Fire," *Decatur Daily*, 6 August 1935, 1; Hugh Sparrow, "Poll Tax Bill Is Kept Alive," *Birmingham Age–Herald*, 7 August 1935, 4; "Governor Says He Will Veto Sterilization," *Catholic Week* (Birmingham), 1 September 1935, 1; William L. Truby, "Today in Both Houses of the Alabama Legislature," *Decatur Daily*, 4 September 1935, 2; "Foes Run Out on Measure to Cut Phone Tax," *Montgomery Advertiser*, 5 September 1935, 7; and Larson, *Sex, Race, and Science*, 143–45.
54. Mary Gawthorpe to Ryan, 3 February 1934; Ryan to Gawthorpe, 6 February 1934; and Gawthorpe to Ryan, 25 February 1934, Ryan Papers. The Mary E. Gawthorp Papers (TAM 275 [R-7264]) are housed at the Tamiment Library and Robert F. Wagner Archives at New York University; a finding aid is available at http://dlib.nyu.edu/findingaids/html/tamwag/gawthorpe.html.
55. Marion S. Olden, "Present Status of Sterilization Legislation in the United States," *Eugenical News* 31, no. 1 (March 1946) (quotation, 5).

NOTES TO CHAPTER 6

1. Frederick Osborn, minutes of the "Conference on Relation of Eugenics and the Church, 1939," 8 May 1939, p. 2, Frederick Osborn Papers I (a three-volume subset of the American Eugenics Society Papers), American Philosophical Society, Philadelphia [hereafter cited as Osborn Papers I].

2. Ibid.
3. Frederick Osborn, "Notes and Memoranda: The Church Looks at Eugenics," *Eugenical News* 24, no. 2 (June 1939).
4. The full text of Rev. Connell's paper, "The Catholic Church and Eugenics," can be found in the Conference on Relation of Eugenics and the Church, 1939, AES Papers. Connell left Mount St. Alphosus in 1940 to teach moral theology at the Catholic University of America. During his tenure there he became dean of the School of Sacred Theology, served as an associate editor of the *American Ecclesiastical Review*, and became one of the nation's best-known moral theologians (see John J. Delaney, "Connell, Francis J.," in *Dictionary of American Catholic Biography* [Garden City, NY: Doubleday, 1984], 108).
5. Connell, "The Catholic Church and Eugenics," 1–3.
6. Mehler, "History of the American Eugenics Society," 111–19.
7. "Excerpts from letters regarding 'A Eugenics Program for the United States,'" October 1935; and handwritten note, November 1935, Osborn Papers I.
8. Osborn, "Opening Remarks from the Annual Meeting," 5 May 1938, 2, Osborn Papers I.
9. Ibid., 4.
10. Ibid.
11. Reilly, *Surgical Solution*, 129.
12. Kevles, *In the Name of Eugenics*, 199.
13. On the postwar history of sterilization in the United States, see Rebecca M. Kluchin, *Fit to be Tied: Sterilization and Reproductive Rights in America, 1950–1980* (New Brunswick, NJ: Rutgers University Press, 2009).
14. Jeffrey M. Burns, *American Catholics and the Family Crisis, 1930–1962: An Ideological and Organizational Response* (New York: Garland Publishing, 1988), 128–35.
15. Bishop O'Hara, quoted in Burns, *American Catholics and the Family Crisis*, 133.
16. Elaine Tyler May, *Homeward Bound: American Families in the Cold War Era* (New York: Basic Books, 1988), 38–51 (quotation, 38).
17. Edgar Schmiedeler, *The Threat of American Decline* (Washington, DC: National Catholic Welfare Conference, ca. 1940), 31.
18. Edgar Schmiedeler, *A Holy War against the Enemies of the Home and the Nation* (Washington, DC: National Catholic Welfare Conference, 1943), 35.
19. May, *Homeward Bound*, 10, 24–25, 39, 59 (for marriage and birth statistics).
20. Ibid., 28.
21. See, e.g., Morris Fishbein, *Successful Marriage: An Authoritative Guide to Problems Related to Marriage from the Beginning of Sexual Attraction to Matrimony and the Successful Rearing of a Family* (Garden City, NY: Doubleday, 1947); and Norman Edwin Himes, *Your Marriage: A Guide to Happiness* (New York: Farrar and Rinehart, 1940).
22. Paul Bowman Popenoe, *Modern Marriage: A Handbook* (New York: Macmillan, 1925), vii. See also Popenoe, *Conservation of the Family* (Baltimore: Williams and Wilkens, 1926).
23. For more on Popenoe and the American Institute of Family Relations, see Wendy Kline, *Building a Better Race: Gender, Sexuality, and Eugenics from the Turn of the Century to the Baby Boom* (Berkeley: University of California Press, 2001), 124–56; Molly Ladd-Taylor, "Eugenics, Sterilization, and Modern Marriage in the USA: The Strange Career of Paul Popenoe," *Gender & History* 13, no. 2 (August 2001); and Alexandra Minna Stern, *Eugenic Nation: Faults and Frontiers of Better Breeding in Modern America* (Berkeley: University of California Press, 2005), 150–81.

24. Paul Popenoe to Edgar Schmiedeler, 27 September 1934; and Schmiedeler to Popenoe, 5 October 1934, NCWC Papers.
25. See, e.g., "Fifteenth Meeting of the Eugenics Committee of the United States of America, October 31, 1925—Minutes" (Minutes [1925–1956], AES Papers), in which Madison Grant raises the question of miscegenation; and "Seventh Meeting of the Board of Directors of the American Eugenics Society, February 14, 1929—Minutes" (ibid.), in which Roswell Johnson reports that the legislative committee is working on both a model marriage bill and a separate bill on miscegenation.
26. Paul Blakely, S.J., "Social Diseases and Legislation for Marriage," *America*, 2 July 1938. Blakely's article included a common confusion that was key to the success of many eugenic policy initiatives. In stating that "the disease can be inherited," Blakely made a significant mistake that one of his fellow Jesuits keenly pointed out in a subsequent issue: that disease cannot be inherited, although fetuses can be infected *in utero*, causing an environmental transmission rather than a hereditary taint. Blakely responded that he had "used the term in its popular, not biological, sense," a misstep for which he considered himself "censurable" (see Robert C. Graham, S.J., "Correspondence: Marriage Legislation," *America*, 16 July 1938; and Blakely, "Correspondence: Marriage Legislation," *America*, 23 July 1938).
27. Francis J. Connell, C.SS.R., "Correspondence: Marriage Legislation," *America*, 16 July 1938.
28. Francis J. Connell, C.SS.R, "May the State Forbid Marriage because of Social Disease?," *Ecclesiastical Review* 49, no. 6 (December 1938).
29. See Thomas Vernor Moore, "Marriage and Venereal Infection: I," *Ecclesiastical Review* 50, no. 4 (April 1939); Francis J. Connell, C.SS.R., "Marriage and Venereal Infection: II," *Ecclesiastical Review* 50, no. 4 (April 1939); Francis J. Connell, C.SS.R., "State Legislation on Venereal Diseases," *Ecclesiastical Review*, 50, no. 5 (May 1939); Francis B. Donnelly, S.J., "Compulsory Blood Tests before Marriage: I," *Ecclesiastical Review* 51, no. 1 (July 1939); Francis J. Connell, C.SS.R, "Compulsory Blood Tests Before Marriage: II," *Ecclesiastical Review* 51, no. 1 (July 1939); Thomas Vernor Moore, O.S.B., "The Marriage of Persons Infected with Venereal Disease: I," *Ecclesiastical Review* 53, no. 1 (July 1940): 27–53; and Francis J. Connell, C.SS.R., "The Marriage of Persons Infected with Venereal Disease: II," *Ecclesiastical Review* 53, no. 1 (July 1940).
30. John LaFarge, S.J., *The Race Question and the Negro* (New York: Longmans, Green, 1943), 196.
31. Martha Hodes, "Introduction: Interconnecting and Diverging Narratives," in *Sex, Love, Race: Crossing Boundaries in North American History*, ed. Martha Hodes (New York: New York University Press, 1999.
32. Although Roman Catholicism had been the most populous single Christian denomination in the United States since the 1870s, in real terms, Protestants overwhelmingly outnumbered Catholics. Thus, the clergy were particularly troubled by the problem of "leakage" due to the presence of Protestant settlement houses and social services. This period of transition and flux prompted such investigations as Gerald Shaughnessy's *Has the Immigrant Kept the Faith?* (New York, 1925).
33. Working from a variety of sources, in 1934 Peter Bernarding estimated that of those persons nationwide involved in mixed marriages, 38 percent, or 79,800 persons, failed to make their Easter duty, hence rendering them "lost to the Church." Furthermore, Bernarding explained, "my constant endeavor in making this estimate has been to underestimate rather than to overestimate; so that I think it safe to set down

our annual losses from this source as being in the neighborhood of 100,000 souls" (Peter Bernarding, A.M., S.T.B., "Catholic Losses through Mixed Marriages," *Homiletic and Pastoral Review* 34 [September 1934]; and Bernarding, "Mixed Marriages: Preventatives and Curatives," *Homiletic and Pastoral Review* 35 [October 1934]).

34. Joseph P. Donovan, C.M., J.C.D., "Keeping Back the Color Line," *Homiletic and Pastoral Review* 31 (May 1931), 812.

35. Of the impediments to marriage recognized by the church, one of the most common was marriage between a Catholic and a non-Catholic. The Code of Canon Law established two types of impediments for these marriages, depending on whether or not the non-Catholic partner had been baptized. If that partner was unbaptized (a Jew, a Muslim, a Hindu, etc.), then there was a "diriment" impediment rendering the marriage invalid. If that partner had been baptized (a Protestant), then there was an "impedient" impediment rendering the marriage valid but gravely sinful. Though this teaching on marriage traced back to the Apostolic period, it had undergone a number of slight alterations, and by the nineteenth century, bishops, including those in America, were allowed to issue a dispensation from the impediment on the condition that the priest secure written promises from the parties that the Catholic partner would be free to practice his or her faith and that the children would be raised and educated in the Catholic faith. For more specific information, see T. Lincoln Bouscaren, S.J., and Adam C. Ellis, S.J., *Canon Law: A Text and Commentary* (Milwaukee: Bruce Publishing, 1946), 455–96; Stanislaus Woywood, O.F.M., LL.B., "Canon Law Studies: Marriages between Catholics and Non-Catholics," *Homiletic and Pastoral Review* 40 (February 1940); Woywood, "Marriage Impediment of Mixed Religion," *Homiletic and Pastoral Review* 40 (March 1940); and Woywood, "Marriage of Catholics to Unbaptized Persons," *Homiletic and Pastoral Review* 40 (April 1940).

36. Peggy Pascoe, *What Comes Naturally: Miscegenation Law and the Making of Race in America* (New York: Oxford University Press, 2009), 1–130; and Williamson, *New People*, 91–109.

37. "The Racial Integrity Act," S. B. 219, Commonwealth of Virginia, passed March 20, 1924. For more on the act, see Barbara Bair, "Remapping the Black/White Body: Sexuality, Nationalism, and Biracial Antimiscegenation Activism in 1920s Virginia," in *Sex, Race, Love: Crossing Boundaries in North American History*, ed. Martha Hodes (New York: New York University Press, 1999); Pascoe, *What Comes Naturally*, 140–50; and Richard B. Sherman, "'The Last Stand': The Fight for Racial Integrity in Virginia in the 1920s," *Journal of Southern History*, 54, no. 1 (February 1988).

38. W. A. Plecker folder, Davenport Papers; Minutes (1925–1956), folders 1 and 2 (1925–1935), AES Papers.

39. [Davenport,] "White America." A copy of the review is also located in the Davenport Papers.

40. Charles B. Davenport and Morris Steggerda, *Race Crossing in Jamaica* (Washington DC: Carnegie Institution of Washington, 1929); Barkan, *Retreat of Scientific Racism*, 162–68.

41. Iris B. Buaken, "You Can't Marry a Filipino: Not If You Live in California," *Commonweal*, 16 March 1945.

42. David W. Southern, "But Think of the Kids: Catholic Interracialists and the Great American Taboo of Race Mixing," *U.S. Catholic Historian* 16, no. 3 (Summer 1998).

43. López, *White by Law*, 9–10.

44. Ibid., 121, 124.

45. Pascoe, *What Comes Naturally*, 163–93.
46. Roger Traynor Majority Opinion, Perez v. Lippold, L.A. 20305, Supreme Court Case Files, California State Archives, Sacramento, CA.
47. Peggy Pascoe, "Miscegenation Law, Court Cases, and Ideologies of 'Race' in Twentieth-Century America," *Journal of American History* 83, no. 1 (June, 1996), 61.
48. Copy of letter from Daniel Marshall to the Most Reverend Joseph T. McGucken, 23 April 1947, John LaFarge, S.J., Papers, Special Collections, Georgetown University Library [hereafter cited as LaFarge Papers].
49. Copy of letter from McGucken to Marshall, 26 April 1947, box 17, folder 29, Daniel Marshall Correspondence, 1947–1950, LaFarge Papers.
50. Petitioner's Reply Brief, Perez v. Moroney [Lippold], L.A. 20305 (8 November 1947), 20–22 (quotation at 21 is originally from the decision in State v. Tutty, 41 Fed. 735 [1890]).
51. Ibid., 23–24.
52. Ibid., 33–44.
53. Barkan, *Retreat of Scientific Racism*.
54. LaFarge, *Race Question and the Negro*, 195. A "new edition expanded and rewritten" of this work was originally published in 1937 under the title *Interracial Justice*.
55. Matthew Frye Jacobson provides an insightful account of this shift from biological determinism to cultural determinism in the second quarter of the twentieth century in his monograph, *Whiteness of a Different Culture* (see pp. 96–109); see also Pascoe, "Miscegenation Law, Court Cases, and Ideologies of 'Race' in Twentieth-Century America."
56. Marshall to LaFarge, 19 September 1947, box 17, folder 29, Daniel Marshall Correspondence, 1947–1950, LaFarge Papers.
57. LaFarge to Marshall, 26 September 1947, box 17, folder 29, Daniel Marshall Correspondence, 1947–1950, LaFarge Papers.
58. LaFarge to Marshall, 26 September 1947, box 17, folder 29, Daniel Marshall Correspondence, 1947–1950, LaFarge Papers.
59. LaFarge to Marshall, 24 November 1947, box 17, folder 29, Daniel Marshall Correspondence, 1947–1950, LaFarge Papers.
60. Petitioner's Reply Brief, Perez v. Moroney [Lippold], L.A. 20305 (8 November 1947), 45–54 (quotation, 48).
61. Shenk, Dissenting Opinion, Perez v. Lippold, L.A. 20305, Supreme Court Case Files, California State Archives, Sacramento.
62. Traynor, Majority Opinion, Perez v. Lippold, L.A. 20305, p. 2, Supreme Court Case Files, California State Archives, Sacramento.
63. Ibid., 22–23.
64. J. Carter, Concurring Opinion, Perez v. Lippold, L.A. 20305, Supreme Court Case Files, California State Archives, Sacramento.
65. Edmonds, Concurring Opinion, Perez v. Lippold, L.A. 20305, Supreme Court Case Files, California State Archives, Sacramento.
66. "Seventy-Six-Year-Old Miscegenation Statute," *The Nation*, 16 October 1948, 415.
67. "Interracial Marriage," *America*, 16 October 1948.
68. Robert F. Drinan, S.J., "Triumph Over Racism," *America*, 22 January 1949.
69. Ted LeBerthon, "Correspondence: Interracial Marriages," *America*, 12 March 1949.
70. "The California Marriage Decision," *Interracial Review* 22, no. 1 (January 1949.
71. For evidence of this shifting position, see Southern, "But Think of the Kids," 83–93.

NOTES TO EPILOGUE

1. Arnaldo Cortesi, "Pope Says Church Raises No Bars against Quest for Scientific Truth," *New York Times*, 9 September 1953).
2. James D. Watson and Francis Crick, "A Structure for Deoxyribose Nucleic Acid," *Nature*, 25 April 1953.
3. Evelyn Fox Keller, *The Century of the Gene* (Cambridge, MA: Harvard University Press, 2000), 22–25.
4. Matthew Frye Jacobson, *Whiteness of a Different Color: European Immigrants and the Alchemy of Race* (Cambridge, MA: Harvard University Press, 1998), 98–105; Barkan, *Retreat of Scientific Racism*, 279–340.
5. Will Herberg, *Protestant—Catholic—Jew: An Essay in American Religious Sociology*, rev. ed. (Garden City, NY: Anchor Books, 1960), 37–39, 151–52.
6. Mark S. Massa, *Catholics and American Culture: Fulton Sheen, Dorothy Day, and the Notre Dame Football Team* (New York: Crossroads, 1999), 1–10.
7. Johanna Schoen, *Choice and Coercion: Birth Control, Sterilization, and Abortion in Public Health and Welfare* (Chapel Hill: University of North Carolina, 2005).
8. Kluchin, *Fit to be Tied*.
9. Reilly, *Surgical Solution*, 151–52. See also May, *Barren in the Promised Land*, 118–25.
10. Kim Severson, "Thousands Sterilized, a State Weighs Restitution," *New York Times*, 9 December 2011, http://www.nytimes.com/2011/12/10/us/redress-weighed-for-forced-sterilizations-in-north-carolina.html.
11. Quoted in "Vatican Paper Publishes a 1975 Condemnation of Direct Sterilization," *New York Times* (12 December 1976).
12. "Unjustified Sterilization," *America*, 14 May 1927.
13. Pius XI, *Casti Connubii* (Encyclical of Pope Pius XI on Christian Marriage . . . , 31 December 1930), par. 71, http://www.vatican.va/holy_father/pius_xi/encyclicals/documents/hf_p-xi_enc_31121930_casti-connubii_en.html.

BIBLIOGRAPHY

ARCHIVAL COLLECTIONS

American Eugenics Society Papers. American Philosophical Society, Philadelphia.

Association for Voluntary Sterilization Papers. Social Welfare History Archives, University of Minnesota.

Cooper, John Montgomery, Papers. Archives of the Catholic University of America, Washington, DC.

Davenport, Charles B., Papers. American Philosophical Society, Philadelphia.

LaFarge, John, S.J., Papers. Special Collections, Georgetown University Library.

National Catholic Welfare Conference Papers. Collections 121 and 121D. Archives of the Catholic University of America, Washington, DC.

Osborne, Frederick, Papers I (a three-volume subset of the American Eugenics Society Papers). American Philosophical Society, Philadelphia.

Ryan, John A., Papers. Archives of the Catholic University of America, Washington, DC.

Schrembs, Joseph, Papers. Archives of the Diocese of Cleveland, Cleveland, OH.

Supreme Court Case Files. California State Archives, Sacramento.

United States Catholic Council, Department of Immigration Records. Center for Migration Studies, Staten Island, NY.

PUBLISHED PRIMARY SOURCES

"The Advisory Council of the Eugenics Committee of the United States of America." *Eugenical News* 8, no. 4 (April 1923): 29.

Alexander, Jack. "A Vast Program Going Nowhere." *America*, 12 March 1932.

American Neurological Association, Committee for the Investigation of Eugenical Sterilization. *Eugenical Sterilization: A Reorientation of the Problem*. New York: Macmillan, 1936.

The American Social Hygiene Association, 1914–1916. New York: American Social Hygiene Assn., 1916.

"A New Issue." *Catholic Universe Bulletin*, 6 May 1927.

"Attitude of Church on Sterilization Reaffirmed by Puerto Rican Prelate." *Catholic Action*, May 1934.

Baker, J. N. "Medical and Health Legislation in 1935." *Journal of the Medical Association of the State of Alabama* 5 (October 1935): 158.

Bernarding, Peter, A.M., S.T.B. "Catholic Losses through Mixed Marriages." *Homiletic and Pastoral Review* 34 (September 1934): 1267–72.

———. "Mixed Marriages: Preventatives and Curratives." *Homiletic and Pastoral Review* 35 (October 1934): 52–58.

Bernstein, Charles, M.D. *Social Care of the Mentally Deficient*. Problems of Mental Deficiency. Washington, DC: National Catholic Welfare Conference, 1930.

Betten, Francis S., S.J. "Sterilization Banned by Natural Law, Says Jesuit." *Catholic Universe Bulletin*, 20 May 1927.

"Bettering the Stock." *America*, 6 October 1934.

"Bishop Byrne Denounces 'Eugenic' Laws Adopted by Puerto Rican Assembly." *Catholic Action*, September 1937.

"Bishops Bring United Counsel to Problems of National Importance." *N.C.W.C. Bulletin* 2 (October 1920): 1.

Blake, Jerome. "Hysterilization." *America*, 9 June 1934.

Blakely, Paul L, S.J. "Correspondence: Marriage Legislation." *America*, 23 July 1938.

———. "Nams, Kallikaks, and Jukeses." *America*, 28 March 1936.

———. "Social Diseases and Legislation for Marriage." *America*, 2 July 1938.

———. "Sterilization and Other Conjectures." *America*, 16 March 1935.

Blanshard, Paul. *American Freedom and Catholic Power*. Boston: Beacon Press, 1949.

———. "Catholic Church and Education." *The Nation*, 15 November 1947.

———. "Catholic Church in Medicine." *The Nation*, 1 November 1947.

———. *Communism, Democracy, and Catholic Power*. Boston: Beacon Press, 1951.

———. *My Catholic Critics*. Boston: Beacon Press, 1952.

———. "Sexual Code of the Roman Church." *The Nation*, 8 November 1947.

"Blanshard Assails Ban on the Nation." *New York Times*, 10 January 1948.

"Blanshard Scores Church 'Censors.'" *New York Times*, 27 June 1948.

Boas, Franz. *Anthropology and Modern Life*. New York: Norton, 1928.

Bolton, Mary V. "Catholic Welfare Bureau: The Care of Mentally Defective Children." *The Tidings*, 22 January 1926.

Bouscaren, T. Lincoln, S.J., and Adam C. Ellis, S.J. *Canon Law: A Text and Commentary*. Milwaukee: Bruce Publishing, 1946.

Bruehl, Charles. *Birth Control and Eugenics in Light of Fundamental Ethical Principles*. New York: Joseph F. Wagner, 1928.

———. "The Church and Race Improvement." *Homiletic and Pastoral Review* 26 (June 1926): 903–13.

———. "The Church and True Eugenics." *Homiletic and Pastoral Review* 27 (April 1927): 689–97.

———. "Eugenical Education." *Homiletic and Pastoral Review* 27 (March 1927): 569–76.

———. "Eugenical Sterilization." *Homiletic and Pastoral Review* 26 (August 1926): 1123–31.

———. "Eugenics and Marriage." *Homiletic and Pastoral Review* 27 (February 1927): 457–65.

———. "Eugenics in the Christian Sense." *Homiletic and Pastoral Review* 27 (May 1927): 801–9.

———. "Moral Aspects of Eugenics." *Homiletic and Pastoral Review* 26 (July 1926): 1011–20.

———. "Moral Aspects of Sterilization." *Homiletic and Pastoral Review* 26 (September 1926): 1235–43.

———. "The Morality of Sterilization." *Homiletic and Pastoral Review* 26 (November 1926): 113–19.

———. "Practical Objections against Legalized Sterilization." *Homiletic and Pastoral Review* 27 (January 1927): 341–48.

———. "The State and Eugenical Sterilization." *Homiletic and Pastoral Review* 26 (October 1926): 1–9.

———. "Sterilization and Heredity." *Homiletic and Pastoral Review* 27 (December 1926): 225–31.

"Brush Foundation Deems Humans Barnyard Animals, Bishop Says." *Catholic Universe Bulletin*, 22 June 1928

Buaken, Iris B. "You Can't Marry a Filipino: Not If You Live in California." *Commonweal*, 16 March 1945.

Buck v. Bell, 274 U.S. 200, 47 S. Ct. 584 (1927).

"The California Marriage Decision." *Interracial Review* 22, no. 1 (January 1949): 4.

"A Catholic Answer." *The Nation*, 3 January 1948.

Chase, John W. "Expanded Articles: American Freedom and Catholic Power." *New York Times*, 15 May 1949, BR15.

"Combating Sterilization Bills." *Central-Blatt and Social Justice* 24 (April 1931): 25–26.

Committee on Special War Activities. *Social Reconstruction: A General Review of Problems and a Survey of Remedies*. Washington, DC: National Catholics War Council, 1919.

Conklin, Edwin Grant. *Heredity and Environment in the Development of Men*. Princeton, NJ: Princeton University Press, 1915.

———. "Some Recent Criticisms of Eugenics." *Eugenical News* 13, no. 5 (May 1928): 61–65.

Connell, Francis J., C.SS.R. "Compulsory Blood Tests before Marriage: II." *Ecclesiastical Review* 51, no. 1 (July 1939): 21–30.

———. "Correspondence: Marriage Legislation." *America*, 16 July 1938.

———. "Marriage and Venereal Infection: II." *Ecclesiastical Review* 50, no. 4 (April 1939): 331–34.

———. "The Marriage of Persons Infected with Venereal Disease: II." *Ecclesiastical Review* 53, no. 1 (July 1940): 54–59.

———. "May the State Forbid Marriage because of Social Disease?" *Ecclesiastical Review* 49, no. 6 (December 1938): 507–18.

———. Review of *American Freedom and Catholic Power*, by Paul Blanchard. *Cornell Law Quarterly* 35 (1949–50): 678–84.

———. "State Legislation on Venereal Diseases." *Ecclesiastical Review* 50, no. 5 (May 1939): 445–46.

Conway, Bertrand L. "The Church and Eugenics." *Catholic World*, November 1928.

Coogan, John E, S.J. "Eugenists in Virginia Betray Science and Democracy." *America*, 16 January 1943.

Cooper, John Montgomery. *Birth Control*. Washington, DC: National Catholic Welfare Council, 1923.

———. "The Birth Rate of Genius: Does Contraception Curb It?," *Eugenics* 2, no. 5 (May 1929): 18.

———. *Religion Outlines for Colleges*. 4 vols. Washington, DC: Catholic Education Press, 1924–30.

"Copy of Encyclical for Each of 1,000 Families in Parish." *The Tidings*, 30 January 1931.

Cortesi, Arnaldo. "Pope Says Church Raises No Bars against Quest for Scientific Truth." *New York Times*, 9 September 1953, 18.

Cox, Ignatius W., S.J. "The Folly of Human Sterilization." *Scientific American*, October 1934.

Cox, Ignatius W., S.J., and James J. Walsh. *What about Sterilization?* Washington, DC: National Catholic Welfare Conference, 1934.

Davenport, Charles B. *Heredity in Relation to Eugenics*. New York: Henry Holt, 1911.

———. "How Probation Works." *Eugenical News* 12, no. 9 (September 1927): 132.

———. "White America." *Eugenical News* 9, no. 1 (January 1924): 3.

Davenport, Charles B., and Morris Steggarda. *Race Crossing in Jamaica*. Washington, DC: Carnegie Institution of Washington, 1929.

Donnelly, Francis B., S.J. "Compulsory Blood Tests before Marriage: I." *Ecclesiastical Review* 51, no. 1 (July 1939): 9–21.

———. "Have You an American Intelligence?" *America*, 14 July 1923.

Donovan, Joseph P., C.M., J.C.D. "Keeping Back the Color Line." *Homiletic and Pastoral Review* 31 (May 1931).

Donovan, Stephen M. "Circa Liceitatem Cujusdam Operationis Chirurgoae Proponuntur Dubia Nonnulla." *American Ecclesiastical Review* 42, no. 9 (March 1910): 271–75.

———. "The Morality of the Operation of Vasectomy." *American Ecclesiastical Review* 44, no. 11 (May 1911): 571–74.

———. "Summing up the Discussion on Vasectomy." *American Ecclesiastical Review* 45, no. 3 (September 1911): 313–19.

Draper, John William. *History of the Conflict between Religion and Science*. New York: D. Appleton and Co., 1897.

Drinan, Robert F., S.J. "Triumph over Racism." *America*, 22 January 1949.

East, Edward M. "The Birth Rate of Genius: Does Contraception Curb It?" *Eugenics* 2, no. 5 (May 1929): 19.

"Encyclical of Pope to be Broadcast." *The Tidings*, 30 January 1931.

"Eugenics: True and False." *Catholic Universe Bulletin*, 6 July 1928.

"Expect Pope to Tighten Rules on Marriage." *Chicago Daily Tribune*, 5 January 1931, 1.

"Father Obering Attacks Alabama Sterilization Law in Birmingham." *The SpringHillian*, 29 April 1935, 1.

"Feminism and the 'New Woman': Their Eugenical Significances." *Eugenics* 3, no. 8 (August 1930).

Fishbein, Morris. *Successful Marriage: An Authoritative Guide to Problems Related to Marriage from the Beginning of Sexual Attraction to Matrimony and the Successful Rearing of a Family*. Garden City, NY: Doubleday, 1947.

"Foes Run Out on Measure to Cut Phone Tax." *Montgomery Advertiser*, 5 September 1935, 7.

"A Free Hand in the Philippines." *Eugenical News* 15, no. 5 (May 1930): 70.

Galton, Francis. *Hereditary Genius: An Inquiry into Its Laws and Consequences*. London, 1869.

———. *Natural Inheritance*. London, 1889.

Gerrard, Thomas J. "Eugenics and Catholic Teaching." *Catholic World*, June 1912.

Gest, John B. "Legal Status of Sterilization Laws in the United States." *American Ecclesiastical Review* 91, no. 4 (October 1934): 401–11.

Gibbons, James Cardinal. "The Pastoral Letter of 1919." In *The National Pastorals of the American Hierarchy, 1792–1919* edited by Peter Guilday, 266–340. Washington DC: National Catholic Welfare Council, 1923.

Goddard, Henry H. *Feeble-Mindedness: Its Causes and Consequences*. New York: Macmillan, 1914.

———. *The Kallikak Family: A Study in the Heredity of Feeble-Mindedness*. New York: Macmillan, 1912.

Gordy, Frank. "House Spotlight Is Taken by Sterilization Measure." *Mobile Register*, 11 May 1935, 1.

Gosney, Ezra S. "Eugenic Sterilization: Human Betterment Demands It." *Scientific American*, July 1934.

Gosney, Ezra S., and Paul Bowman Popenoe, eds. *Sterilization for Human Betterment: A Summary of Results of 6,000 Operations in California, 1909–1929*. New York: Macmillan, 1929.

"Governor Says He Will Veto Sterilization." *Catholic Week* (Birmingham), 1 September 1935, 1.

Graham, Robert C., S.J. "Correspondence: Marriage Legislation." *America*, 16 July 1938.

Grant, Madison. *The Passing of the Great Race*. New York: Charles Scribner's Sons, 1916.

"Grass-Roots Eugenics." *Commonweal*, 16 July 2004.

"The House Passes the Sterilization Bill." *Birmingham News*, 11 May 1935, 4.

"How Old Are You?" *America*, 22 May 1926.

"Human Rights in Germany." *America*, 6 January 1934.

"Interracial Marriage." *America*, 16 October 1948.

"In the Field of Immigration." *N.C.W.C. Bulletin* 7 (June 1925): 9.

"Is Christian Morality Harmful? Over-Charitable to the Unfit?" *Eugenics* 1, no. 3 (December 1928): 20–21.

"Is Eugenics Racial Snobbery?" *Eugenics* 2, no. 2 (February 1929): 20–21.

Harris, Reginald G. "Eugenics in South America." *Eugenical News* 7, no. 3 (March 1922): 29–30.

Hauber, Ulrich A. *Inheritance of Mental Defect*. Problems of Mental Deficiency. Washington, DC: National Catholic Welfare Conference, 1930.

Hawley, Hudson. "Pope Pius Will Issue Encyclical against Birth Control Spread." *Atlanta Constitution*, 5 January 1931, 1.

Herberg, Will. *Protestant—Catholic—Jew: An Essay in American Religious Sociology*. Rev. ed. Garden City, NY: Anchor Books, 1960.

Himes, Norman Edwin. *Your Marriage: A Guide to Happiness*. New Your: Farrar and Rinehart, 1940.

Historical Census Browser. University of Virginia, Geospatial and Statistical Data Center. 2004. http://mapserver.lib.virginia.edu/.

Holmes, Samuel J. "The Catholic View." *Eugenics* 1, no. 2 (November 1928): 37–38.

Hurst, Ralph. "$2000 Tax-Free Homesteads Is Urged in House." *Birmingham News*, 10 May 1935, 1.

Jennings, Herbert Spencer. *The Biological Basis of Human Nature*. London: Faber and Faber, 1930.

———. "Undesirable Aliens." *The Survey* 51 (1923): 309–12, 364.

Johnson, Roswell. "Legislation." *Eugenics* 4, no. 2 (February 1931): 73.

"Judge Orders Couple Not to Conceive." A.P. Wire Service, 8 May 2004.

Kluchin, Rebecca M. *Fit to be Tied: Sterilization and Reproductive Rights in America, 1950–1980*. New Brunswick, NJ: Rutgers University Press, 2009.

Labouré, Theodore. "Is Vasectomy Then Unlawful?" *American Ecclesiastical Review* 44, no. 11 (May 1911): 574–83.

———. "One Last Remark on Vasectomy." *American Ecclesiastical Review* 45, no. 3 (September 1911): 355–59.

LaFarge, John, S.J. *The Race Question and the Negro*. New York: Longmans, Green, 1943.

Landman, Jacob H. *Human Sterilization*. New York: Macmillan, 1932.

———. "Race Betterment by Human Sterilization." *Scientific American*, June 1934.

Largent, Mark. *Breeding Contempt: The History of Coerced Sterilization in the United States*. New Brunswick, NJ: Rutgers University Press, 2007.

Laughlin, Harry H. *Analysis of America's Modern Melting Pot: Hearings Before the House Committee on Immigration and Naturalization, House of Representatives, Sixty-Seventh Congress, Third Session. November 21, 1922. Serial 7-C*. Washington, DC: Government Printing Office, 1923.

———. *Eugenical Sterilization: 1926: Historical, Legal, and Statistical Review of Eugenical Sterilization in the United States*. New Haven, CT: American Eugenics Society, 1926.

———. *Eugenical Sterilization in the United States*. Chicago: Psychopathic Laboratory of the Municipal Court of Chicago, 1922.

———. *The Legal Status of Eugenical Sterilization*. Chicago: Psychopathic Laboratory of the Municipal Court of Chicago, 1929.

LeBerthon, Ted. "Correspondence: Interracial Marriages." *America*, 12 March 1949.

Lehane, Joseph B. *The Morality of American Civil Legislation Concerning Eugenical Sterilization*. Washington, DC: Catholic University of America Press, 1944.

Leo XIII. *Aeterni Patris*. Encyclical of Pope Leo XIII on the Restoration of Christian Philosophy. 4 August 1879. http://www.vatican.va/holy_father/leo_xiii/encyclicals/documents/hf_l-xiii_enc_04081879_aeterni-patris_en.html.

———. *Rerum Novarum*. Encyclical of Pope Leo XIII on Capital and Labor. 15 May 1891. http://www.vatican.va/holy_father/leo_xiii/encyclicals/documents/hf_l-xiii_enc_15051891_rerum-novarum_en.html.

"Letters on the Catholic Series." *The Nation*, 13 December 1947.

"Local Minister Praises, Criticizes Teachings of Pope in Encyclical." *Atlanta Constitution*, 12 January 1931, 7.

MacArthur, Kenneth C. "Eugenics and the Church." *Eugenics* 4, no. 2 (February 1931): 74.

"Man's Dignity." *America*, 13 June 1942.

Mayer, Joseph, S.J.. "Eugenics in Roman Catholic Literature." *Eugenics* 3, no. 2 (February 1930): 43–51.

———. *Gesetlziche Unfruchtbarmachung Geisteskranker*. Freiburg im Breisgau: Herder and Co., 1927.

McGowan, R. A. "The New Immigration Bill." *The Catholic Charities Review* 8, no. 2 (February 1924): 58.

McGurn, Barrett. "Three Vicious Laws Effective in Puerto Rico: An 'Easy' Way of Killing an Excess Population." *America*, 24 July 1937.

McKenna, Daniel J. "'Science,' the Ally of Bigotry." *America*, 4 December 1926.

McSweeney, Chris J. *The Case against Sterilization of Mental Defectives*. St. Louis: Central Bureau, Catholic Central Verein of America, 1931.

"Modern Marital Evils." *Catholic Universe Bulletin*, 16 January 1931, 1.

Montavon, William F. *Eugenic Sterilization in the Laws of the States*. Problems of Mental Deficiency. Washington, DC: National Catholic Welfare Conference, 1930.

Moore, Thomas Vernor, O.S.B. "Marriage and Venereal Infection: I." *Ecclesiastical Review* 50, no. 4 (April 1939): 323–31.

———. "The Marriage of Persons Infected with Venereal Disease: I." *Ecclesiastical Review* 53, no. 1 (July 1940): 27–53.

———. "Types and Causes of Feeble-Mindedness." *Proceedings of the Fourth National Conference of Catholic Charities*, 231–51. Washington, DC: National Conference of Catholic Charities, 1916.

"The Morality and Lawfulness of Vasectomy." *American Ecclesiastical Review* 44, no. 11 (May 1911): 562–83.

"The Morality and Lawfulness of Vasectomy." *American Ecclesiastical Review* 45, no. 1 (July 1911): 71–77.

Murphy, John C. *An Analysis of the Attitudes of American Catholics toward the Immigrant and the Negro, 1825–1925*. Washington, DC: Catholic University of America, 1940.

National Council of Catholic Men. *Special Bulletin on Ohio Sterilization and Marriage License Bills*. Washington, DC: National Catholic Welfare Conference, 1927.

NCWC News Service. "Quantity Production Marks Output in Legislatures of Many States." *The Tidings*, 25 March 1927.

———. "State Legislatures Discuss Anti-evolution Measures Propose Marriage Reforms." *The Tidings*, 25 February 1927.

———. "State Legislatures Discuss Bills of Interest to Catholics." *The Tidings*, 4 March 1927.

———. "State Legislatures Struggle with Several Eugenic Measures." *The Tidings*, 22 April 1927.

"The New Conquest of California." *Eugenical News* 13, no. 2 (February 1928).

"The New Immigration Law," *Catholic Charities Review* 1 (April 1917): 101.

"News and Notes." *Eugenical News* 10, no. 3 (March 1925): 36.

Obering, William F., S.J. "Authority Says Alabama Sterilization Measure Is Declaration of Slavery." *Catholic Week* (Birmingham), 5 May 1935.

O'Keefe, Ruth E. "We Catholics Have a Duty." *The Nation*, 27 May 1939.

Olden, Marion S. "Present Status of Sterilization Legislation in the United States." *Eugenical News* 31, no. 1 (March 1946): 3–14.

———. *Sterilization and the Organized Opposition*. Princeton, NJ: Sterilization League of New Jersey, 1936.

O'Malley, Austin. "Vasectomy in Defectives." *American Ecclesiastical Review* 44, no. 12 (June 1911): 684–705.

Osborn, Fredrick. "Notes and Memoranda: The Church Looks at Eugenics." *Eugenical News* 24, no. 2 (June 1939): 25.

———. *Preface to Eugenics*. New York: Harper and Brothers, 1940.

Osgood, Phillips E. "The Refiner's Fire." *Eugenics* 1, no. 3 (December 1928): 10–15.

"Our Common Catholic Interests." *Catholic Action*, July 1935.

Parsons, Wilfrid, S.J. "Sterilization Is Criminal Folly!" *America*, 10 February 1934.

Pearl, Raymond. "The Biology of Superiority." *American Mercury*, November 1927.

Perez v. Lippold, 198 P. 2d 17 (1948).

"Personal Factors in Immigration: Should They Get More Attention?" *Eugenics* 2, no. 7 (July 1929): 18–20.

"The Pilgrim." *America*, 6 July 1935.

Pius X. *Pascendi Dominici Gregis*. Encyclical of Pope Pius X on the Doctrine of the Modernists. 8 September 1907. http://www.vatican.va/holy_father/pius_x/encyclicals/documents/hf_p-x_enc_19070908_pascendi-dominici-gregis_en.html.

Pius XI. *Casti Connubii*. Encyclical of Pope Pius XI on Christian Marriage. . . . 31 December 1930. http://www.vatican.va/holy_father/pius_xi/encyclicals/documents/hf_p-xi_enc_31121930_casti-connubii_en.html.

"Pontiff Proclaims Sanctity of Matrimony." *Los Angeles Times*, 9 January 1931, 1.

Popenoe, Paul Bowman. "A Debate on Sterilization: Intelligent Eugenics." *The Forum*, July 1935.

———. *Conservation of the Family*. Baltimore: Williams and Wilkens, 1926.

———. *Marriage Is What You Make It* (1950). New York: Macmillan, 1959.

———. *Modern Marriage: A Handbook*. New York: Macmillan, 1925.

———. "A Roman Catholic View of Sterilization." *Journal of Social Hygiene* 14, no. 3 (March 1928), 186–88. Reprinted in *Sterilization for Human Betterment*, edited by E. S. Gosney and Paul Popenoe (New York: Macmillan, 1929), 186–91.

Popenoe, Paul Bowman, and Roswell Johnson. *Applied Eugenics*. Rev. ed. New York: Macmillan, 1933.

"Pope Pius XI's Encyclical on Marriage." *Eugenical News* 16, no. 2 (February 1931): 21.

"Pope Strict with Cupid." *Los Angeles Times*, 5 January 1931, 1.

"Prizes for Sermons on Eugenics." *Eugenical News* 11, no. 3 (March 1926): 48.

"P. Schmitt on the Gravity of the Mutilation Involved in Vasectomy." *American Ecclesiastical Review* 45, no. 3 (September 1911): 360–62.

"Refusal, Seton Hall College." *Eugenics* 3, no. 4 (April 1930): 155.

"Report of Sub-committee on the Ultimate Program to be Developed by the Eugenics Society of the United States of America." *Eugenical News* 8, no. 8 (August 1923): 73–80.

"The Reproduction Rate of Genius: Will Birth Control Diminish It?" *Eugenics* 2, no. 3 (March 1929): 22–24.

"Resolution 15: The Life and Witness of the Christian Community—Marriage and Sex." Lambeth Conference, Resolutions from 1930. http://www.lambethconference.org/resolutions/1930/1930-15.cfm.

"Rome Has Spoken!" *The Christian Century*, 4 February 1931.

Ross, Edward A. "The Causes of Race Superiority." *Annals of the American Academy of Political and Social Science* 18 (July 1901), 67–89.

Ryan, John A. *The Catholic Church and the Citizen*. New York: Macmillan, 1928.

———. "The Church and Birth Control." In *Family Limitation and The Church and Birth Control*, by John Ryan, 17–23. New York: Paulist Press, 1916.

———. "A Debate on Sterilization: Futile Immorality." *The Forum*, July 1935.

———. "Family Limitation." In *Family Limitation and The Church and Birth Control*. New York: Paulist Press, 1916.

———. *Human Sterilization*. Washington, DC: National Catholic Welfare Conference, 1927.

———. *A Living Wage: Its Ethical and Economic Aspects*. New York: Macmillan, 1906.

———. *Moral Aspects of Sterilization*. Problems of Mental Deficiency. Washington, DC: National Catholic Welfare Conference, 1930.

———. "Moral Teaching of the Encyclical on Christian Marriage." *American Ecclesiastical Review* 84, no. 9 (March 1931): 264–71.

———. *Social Doctrine in Action: A Personal History*. New York: Harper and Brothers, 1941.

———. *The State and the Church*. Washington, DC: National Catholic Welfare Conference, 1922.

"Salutary Words." *Commonweal*, 15 March 1935.

Sanger, Margaret. *Family Limitation*. Rev. 5th ed. Margaret Sanger, 1916.

———. *Margaret Sanger: An Autobiography*. New York: W. W. Norton, 1938.

———. *The Pivot of Civilization*. 1922. Reprint, London: J. Cape, 1923.

———. *Woman and the New Race*. New York: Truth Publishing, 1920.

Schmiedeler, Edgar. "Conserving the Family." In *Conserving the Family*, by Edgar Schmiedeler, 19–24. Paterson, NJ: St. Anthony Guild Press, 1932. Originally published in *Catholic Action* (January 1932).

———. *A Holy War against the Enemies of the Home and the Nation*. Washington, DC: National Catholic Welfare Conference, 1943.

———. *Sterilization in the United States*. Washington, DC: National Catholic Welfare Conference, 1943.

———. *The Threat of American Decline*. Washington, DC: National Catholic Welfare Conference, ca. 1940.

"Schmitt's Argument." *American Ecclesiastical Review* 44, no. 11 (May 1911): 565–71.

"The Schools and 'The Nation.'" *New York Times*, 28 May 1949, 14.

"Senate Approves Decision to Boost Income Tax." *Decatur Daily*, 26 June 1935, 1.

"Sermons on Eugenics." *Eugenical News* 11, no. 4 (April 1926): 56.

"Seventy-Six-Year-Old Miscegenation Statute." *The Nation*, 16 October 1948.

Severson, Kim. "Thousands Sterilized, a State Weighs Restitution." *New York Times*, 9 December 2011. http://www.nytimes.com/2011/12/10/us/redress-weighed-for-forced-sterilizations-in-north-carolina.html.

Shaughnessy, Gerald. *Has the Immigrant Kept the Faith?* New York, 1925.

Skinner v. Oklahoma, 316 U.S. 535 (1942).

Solons Hear Opposition to Sterilization." *Catholic Week* (Birmingham), 19 May 1935.

Somerville, Henry. "Eugenics and the Feeble-Minded." *Catholic World*, May 1917.

Sparrow, Hugh. "Poll Tax Bill Is Kept Alive." *Birmingham Age–Herald*, 7 August 1935, 4.

———. "Sterilization Bill for State Up to Governor." *Birmingham Age–Herald*, 1 June 1935, 1.

"Sterilization Bill Gets Veto of Governor Graves." *Mobile Register*, 26 June 1935, 1.

"Sterilization Bill Is Killed." *Catholic Universe Bulletin*, 15 April 1927.

"Sterilization Bill Killed by Veto." *Montgomery Advertiser*, 26 June 1935, 1.

"Sterilization Bill Passes, Awaits Governor's Action." *Mobile Register*, 1 June 1935, 1.

"Sterilization Bill Passes Senate." *Montgomery Advertiser*, 1 June 1935, 1, 3.

"Sterilization Condemned as Contrary to the Natural Law." *Catholic Action*, September 1937.

"Sterilization Conflicts." *Commonweal*, 12 April 1935.

"The Sterilization Craze." *America*, 14 March 1931.

"Sterilization in Alabama." *Montgomery Advertiser*, 6 February 1935, 4.

"Sterilization in the United States." *Commonweal*, 20 September 1935.

"Sterilization: Its Legality, Need; A Catholic and Some Eugenicists Speak." *Eugenics* 3, no. 5 (May 1930): 180–81.

"Sterilization Law in Germany." *American Ecclesiastical Review* 95, no. 1 (July 1936): 50–68.

"Sterilization Progresses." *Montgomery Advertiser*, 11 May 1935, 4.

"Study Clubs Are Urged by D.C.C.W." *Catholic Universe Bulletin*, 21 June 1929.

Taylor, Edmond. "Pope Flays Sex Codes of Today as Deadly Sins." *Chicago Daily Tribune*, 9 January 1931, 1.

Thomalla, C., M.D. "The Sterilization Law in Germany." *Scientific American*, September 1934.

Truby, William L. "House Passes Sterilization Bill." *Decatur Daily*, 10 May 1935, 1.

———. "Sedition Bill Veto Is Under Fire." *Decatur Daily*, 6 August 1935, 1.

———. "Today in Both Houses of the Alabama Legislature." *Decatur Daily*, 4 September 1935, 2.

United States. Department of Commerce and Labor. Bureau of the Census. *Census of Religious Bodies: 1926*. Vol. 2, *Separate Denominations: Statistics, History, Doctrine, Organization, and Work*. Washington, DC: Government Printing Office, 1929.

———. *Census of Religious Bodies: 1936*. Vol. 1, *Summary and Detailed Tables*. Washington DC: Government Printing Office, 1941.

———. *Census of Religious Bodies: 1936*. Vol. 2, *Denominations: Statistics, History, Doctrine, Organization, and Work*. Washington DC: Government Printing Office, 1941.

"Unjustified Sterilization." *America*, 14 May 1927.

"Vatican Paper Publishes a 1975 Condemnation of Direct Sterilization." *New York Times*, 12 December 1976, 23.

Walsh, James J. "Race Betterment." *Commonweal*, 2 February 1934.

———. "This Hereditary Delusion." *Extension Magazine*, February 1931.

Ward, Patrick J. "The Grave Issue of Sterilization." *Catholic Action*, February 1934.

Watson, James D., and Francis Crick. "A Structure for Deoxyribose Nucleic Acid." *Nature*, 25 April 1953.

White, Andrew Dickson. *A History of the Warfare of Science with Theology in Christendom*. New York: D. Appleton and Co., 1901.

Whitney, Leon F. *The Case for Sterilization*. New York: Frederick A. Stokes, 1934.

"Who Are the Unfit?" *America*, 22 May 1932.

"Why Study Clubs?" *Catholic Universe Bulletin*, 8 August 1930.

Wickham, Harvey. "Straining the Quality of Mercy." *America*, 4 December 1926.

Woywood, Stanislaus, O.F.M. "Canon Law Studies: Marriages between Catholics and Non-Catholics." *Homiletic and Pastoral Review* 40 (February 1940): 409–19.

———. "Marriage Impediment of Mixed Religion." *Homiletic and Pastoral Review* 40 (March 1940): 633–43.

———. "Marriage of Catholics to Unbaptized Persons." *Homiletic and Pastoral Review* 40 (April 1940): 756–66.

Zahm, John. *Evolution and Dogma*. Chicago: McBride, 1896.

PUBLISHED BOOKS AND ARTICLES

Abell, Aaron I. *American Catholicism and Social Action: A Search for Social Justice, 1865–1950*. Garden City, NY: Hanover House, 1960.

Adams, Mark B., ed. *The Wellborn Science: Eugenics in Germany, France, Brazil, and Russia*. Oxford: Oxford University Press, 1990.

Allen, Garland E. "Davenport, Charles Benedict." In *American National Biography Online*. Oxford University Press, 2000–2010. http://www.anb.org/articles/13/13-00392.html.

Anderson, Benedict. *Imagined Communities: Reflections on the Origins and Spread of Nationalism*. Rev. ed. London: Verso, 1991.

Appleby, R. Scott. *"Church and Age Unite!": The Modernist Impulse in American Catholicism*. Notre Dame, IN: University of Notre Dame Press, 1992.

———. "Exposing Darwin's 'Hidden Agenda': Roman Catholic Responses to Evolution, 1875–1925." In *Disseminating Darwinism: The Role of Place, Race, Religion, and Gender*, , edited by Ronald L. Numbers and John Stenhouse, 173–208. Cambridge: Cambridge University Press, 1999.

Artigas, Mariano, Thomas F. Glick, and Rafael A. Martínez. *Negotiating Darwin: The Vatican Confronts Evolution, 1877–1902*. Baltimore, MD: Johns Hopkins University Press, 2006.

Bair, Barbara. "Remapping the Black/White Body: Sexuality, Nationalism, and Biracial

Antimiscegenation Activism in 1920s Virginia." In *Sex, Race, Love: Crossing Boundaries in North American History*, edited by Martha Hodes, 399–422. New York: New York University Press, 1999.

Barkan, Elazar. *The Retreat of Scientific Racism: Changing Concepts of Race in Britain and the United States between the World Wars*. Cambridge: Cambridge University Press, 1992.

Bederman, Gail. *Manliness and Civilization: A Cultural History of Gender and Race in the United States, 1880–1917*. Chicago: University of Chicago Press, 1995.

Bowler, Peter J. *Darwinism*. New York: Twayne, 1993.

Braham, Randolph L., ed. *The Vatican and the Holocaust: The Catholic Church and the Jews during the Nazi Era*. New York: Columbia University Press, 2000.

Briggs, Laura. *Reproducing Empire: Race, Sex, Science, and U.S. Imperialism in Puerto Rico*. Berkeley: University of California Press, 2002.

Bristow, Nancy K. *Making Men Moral: Social Engineering during the Great War*. New York: New York University Press, 1996.

Broberg, Gunnar, and Nils Roll-Hansen, eds. *Eugenics and the Welfare State: Sterilization Policy in Denmark, Sweden, Norway, and Finland*. East Lansing: Michigan State University Press, 1996.

Broderick, Francis L. *Right Reverend New Dealer: John A. Ryan*. New York: Macmillan, 1963.

Brooke, John Hedley, and Geoffrey N. Cantor. *Reconstructing Nature: The Engagement of Science and Religion*. New York: Oxford University Press, 2000.

Brown, Dorothy M., and Elizabeth McKeown. *The Poor Belong to Us: Catholic Charities and American Welfare*. Cambridge, MA: Harvard University Press, 1997.

Burns, Jeffrey M. *American Catholics and the Family Crisis, 1930–1962: An Ideological and Organizational Response*. New York: Garland Publishing, 1988.

Caplan, Paula J. "Mother-Blaming." In *"Bad" Mothers: The Politics of Blame in Twentieth-Century America*, edited by Molly Ladd-Taylor and Lauri Umansky, 127–44. New York: New York University Press, 1998.

Carey, Allison C. "Gender and Compulsory Sterilization Programs in America: 1907–1950." *Journal of Historical Sociology* 11, no. 1 (March 1998): 74–105.

Catholic Library Association. *The Catholic Periodical Index: A Guide to Catholic Magazines*. Scranton, PA: National Catholic Educational Association, 1930–.

Chase, Allan. *The Legacy of Malthus: The Social Costs of the New Scientific Racism*. Urbana: University of Illinois Press, 1980.

Chesler, Ellen. *Woman of Valor: Margaret Sanger and the Birth Control Movement in America*. New York: Anchor Books, 1992.

Coburn, Carol K., and Martha Smith. *Spirited Lives: How Catholic Nuns Shaped Catholic Culture and American Life, 1836–1920*. Chapel Hill: University of North Carolina Press, 1999.

Cornwell, John. *Hitler's Pope: The Secret History of Pius XII*. New York: Viking, 1999.

Curran, R. Emmett. "Splendid Poverty: Jesuit Slave-Holding in Maryland, 1805–1838." In *Catholics in the Old South: Essays in Church and Culture*, edited by Randall Miller and Jon Waklyn, 125–46. Macon, GA: Mercer University Press, 1983.

Daniel, W. Harrison. "Virginia." In *Religion in the Southern States*, edited by Samuel S. Hill, 335–62. Macon, GA: Mercer University Press, 1983.

Davidson, Keay. "Huntington, Ellsworth." In *American National Biography Online*. Oxford University Press, 2000–2010. http://www.anb.org/articles/13/13-00830.html.

Davis, Cyprian. *The History of Black Catholics in the United States*. New York: Crossroads, 1990.

Davis, F. James. *Who Is Black? One Nation's Definition*. University Park: Penn State University Press, 1991.

Degler, Carl N. *At Odds: Women and the Family in America from the Revolution to the Present*. Oxford: Oxford University Press, 1980.

———. *In Search of Human Nature: The Decline and Revival of Darwinism in American Social Thought*. New York: Oxford University Press, 1991.

Delaney, John J. "Connell, Francis J." In *Dictionary of American Catholic Biography*, 108. Garden City, NY: Doubleday, 1984.

Dietrich, Donald J. "Catholic Eugenics in Germany, 1920–1945: Hermann Muckermann, S.J. and Joseph Mayer." *Journal of Church and State* 34, no. 3 (Summer 1992): 575–600.

———. "Catholic Resistance to Biological and Racist Eugenics in the Third Reich." In *Germans against Nazism: Nonconformity, Opposition, and Resistance in the Third Reich*, edited by Francis R. Nicosia and Lawrence D. Stokes, 137–55. New York: Berg, 1990.

Dolan, Jay P. *The American Catholic Experience: A History from Colonial Times to the Present*. Notre Dame, IN: University of Notre Dame Press, 1992.

———. *The Immigrant Church: New York's Irish and German Catholics, 1815–1865*. Baltimore: Johns Hopkins University Press, 1975.

Dowbiggin, Ian. "'A Rational Coalition,': Euthanasia, Eugenics, and Birth Control in America, 1940–1970." *Journal of Policy History* 14, no. 3 (2002): 223–60.

Dumenil, Lynn. "The Tribal Twenties: 'Assimilated' Catholics' Response to Anti-Catholicism in the 1920s." *Journal of American Ethnic History* 11, no. 1 (Fall 1991): 21–49.

Durst, Dennis L. "'No Legacy Annuls Heredity from God': Evangelical Social Reformers and the North American Eugenics Movement." PhD diss., Saint Louis University, 2002.

Dyer, Richard. *White*. London: Routledge, 1997.

Dyer, Thomas G. *Theodore Roosevelt and the Idea of Race*. Baton Rouge: Louisiana State University Press, 1980.

Filene, Peter. *Him/Her/Self: Gender Identities in Modern America*. 3rd ed. Baltimore: Johns Hopkins University Press, 1998.

Flynt, Wayne. *Alabama Baptists: Southern Baptists in the Heart of Dixie*. Tuscaloosa: University of Alabama Press, 1998.

Gaillardetz, Richard R. "John A. Ryan: An Early Revisionist." *Journal of Religious Ethics* 18, no. 2 (Summer 1990): 110–17.

Gallagher, Nancy L. *Breeding Better Vermonters: The Eugenics Project in the Green Mountain State*. Hanover, NH: University Press of New England, 1999.

Gerstle, Gary. *American Crucible: Race and Nation in the Twentieth Century*. Princeton, NJ: Princeton University Press, 2001.

Ginsburg, Faye D., and Rayna Rapp, eds. *Conceiving the New World Order: The Global Politics of Reproduction*. Berkeley: University of California Press, 1995.

Gleason, Philip. *Keeping the Faith: American Catholicism Past and Present*. Notre Dame, IN: University of Notre Dame Press, 1987.

Goldhagen, Daniel Jonah. *A Moral Reckoning: The Role of the Catholic Church in the Holocaust and Its Unfulfilled Duty to Repair*. New York: Knopf, 2002.

Gordon, Linda. *Woman's Body, Woman's Right: Birth Control in America*. Rev. ed. New York: Penguin, 1990.

Gossett, Thomas F. *Race: The History of an Idea in America*. 1963. Reprint, New York: Oxford University Press, 1997.

Gould, Stephen Jay. *The Mismeasure of Man*. New York: W. W. Norton, 1981.

Graham, Robert. "The 'Right to Kill' in the Third Reich: Prelude to Genocide." *The Catholic Historical Review* 62, no. 1 (1976): 56–76.

Greeley, Andrew. *The Catholic Myth: The Behaviors and Beliefs of American Catholics*. New York: Scribners, 1990.

Hall, Lesley A. "Women, Feminism, and Eugenics." In *Essays in the History of Eugenics*, edited by Robert A. Peel, 36–51. London: The Galton Institute, 1998.

Haller, Mark H. *Eugenics: Hereditarian Attitudes in American Thought*. New Brunswick, NJ: Rutgers University Press, 1963.

Hasian, Marouf Arif. *The Rhetoric of Eugenics in Anglo-American Thought*. Athens: University of Georgia Press, 1996.

Heineman, Kenneth J. *A Catholic New Deal: Religion and Reform in Depression Pittsburgh*. University Park: Pennsylvania State University Press, 1999.

Higham, John. *Strangers in the Land: Patterns of American Nativism, 1860–1925*. Rev. ed. New York: Atheneum, 1975.

Hodes, Martha. "Introduction: Interconnecting and Diverging Narratives." In *Sex, Love, Race: Crossing Boundaries in North American History*, edited by Martha Hodes, 1–9. New York: New York University Press, 1999.

Hofstadter, Richard. *Social Darwinism in American Thought*. First rev. ed. Boston: Beacon Press, 1955.

Jacobson, Matthew Frye. *Whiteness of a Different Color: European Immigrants and the Alchemy of Race*. Cambridge, MA: Harvard University Press, 1998.

Jenkins, Philip. *The New Anti-Catholicism: The Last Acceptable Prejudice*. Oxford: Oxford University Press, 2003.

Kauffman, Christopher J. *Ministry and Meaning: A Religious History of Catholic Health Care in the United States*. New York: Crossroads Publishing, 1995.

Keller, Evelyn Fox. *The Century of the Gene*. Cambridge, MA: Harvard University Press, 2000.

Kerber, Linda K. "The Republican Mother: Women and the Enlightenment—An American Perspective." In *Toward an Intellectual History of Women*, by Linda K. Kerber, 41–62. Chapel Hill: University of North Carolina Press, 1997.

Kevles, Daniel J. *In the Name of Eugenics: Genetics and the Uses of Human Heredity*. Cambridge, MA: Harvard University Press, 1995.

King, Miriam, and Steven Ruggles. "American Immigration, Fertility, and Race Suicide at the Turn of the Century." *Journal of Interdisciplinary History* 22, no. 3 (Winter 1990): 347–69.

Kinzer, Donald L. *An Episode in Anti-Catholicism: The American Protective Association*. Seattle: University of Washington, 1964.

Kline, Wendy. *Building a Better Race: Gender, Sexuality, and Eugenics from the Turn of the Century to the Baby Boom*. Berkeley: University of California Press, 2001.

Kraut, Alan. *Silent Travelers: Germs, Genes, and the "Immigrant Menace."* New York: Basic Books, 1994.

Kühl, Stefan. *The Nazi Connection: Eugenics, American Racism, and German National Socialism*. New York: Oxford University Press, 1994.

Ladd-Taylor, Molly. "Eugenics, Sterilization, and Modern Marriage in the USA: The Strange Career of Paul Popenoe." *Gender & History* 13, no. 2 (August 2001): 298–327.

Larson, Edward J. *Sex, Race, and Science: Eugenics in the Deep South*. Baltimore: Johns Hopkins University Press, 1995.

Lee, Erika. *At America's Gates: Chinese Immigration during the Exclusion Era, 1882–1943*. Chapel Hill: University of North Carolina Press, 2003.

Leon, Sharon M. *Before "Casti connubii": Early Catholic Responses to the Eugenics Movement in the United States*. Working Papers Series 32, no. 1. [South Bend, IN]: Cushwa Center for the Study of American Catholicism, University of Notre Dame, 2000.

———. "'Hopelessly Entangled in Nordic Pre-suppositions': Catholic Participation in the American Eugenics Society in the 1920s." *Journal of the History of Medicine and Allied Sciences* 59, no. 1 (January 2004): 3–39.

———. "'A Human Being, and Not a Mere Social Factor': Catholic Strategies for Dealing with Sterilization Statutes in the 1920s." *Church History* 73, no. 2 (June 2004): 383–411.

———. "Promoting Wise Marriages: Paul Popenoe, Eugenics, and Marriage Guides for Men in the 1920s." Paper presented at the annual meeting of the American Studies Association, Montreal, ON, Canada, October 29, 1999.

———. "Tensions Not Unlike that Produced by a Mixed Marriage: Daniel Marshall and Catholic Challenges to Interracial Anti-miscegenation Statutes." *U.S. Catholic Historian* 26, no. 1 (December 2008): 27–44.

Lindberg, David C., and Ronald Numbers, eds. *God and Nature: Historical Essays on the Encounter between Christianity and Science*. Berkeley: University of California Press, 1986.

Linkh, Robert M. *American Catholicism and the European Immigrants, 1900–1924*. Staten Island, NY: Center for Migration Studies, 1975.

Lombardo, Paul A. *Three Generations, No Imbeciles: Eugenics, the Supreme Court, and "Buck v. Bell."* Baltimore: Johns Hopkins University Press, 2008.

López, Ian Haney. *White by Law: The Legal Construction of Race*. New York: New York University Press, 1996.

Ludmerer, Kenneth M. *Genetics and American Society: A Historical Appraisal*. Baltimore: Johns Hopkins University Press, 1972.

———. "Genetics, Eugenics, and the Immigration Restriction of 1924." *Bulletin of the History of Medicine* 46, no. 1 (January 1972): 59–81.

Massa, Mark S. *Anti-Catholicism in America: The Last Acceptable Prejudice*. New York: Crossroads, 2003.

———. *Catholics and American Culture: Fulton Sheen, Dorothy Day, and the Notre Dame Football Team*. New York: Crossroads, 1999.

May, Elaine Tyler. *Barren in the Promised Land: Childless Americans and the Pursuit of Happiness*. Cambridge, MA: Harvard University Press, 1995.

———. *Homeward Bound: American Families in the Cold War*. New York: Basic Books, 1988.

McGreevy, John T. *Catholicism and American Freedom: A History*. New York: W. W. Norton, 2003.

———. *Parish Boundaries: The Catholic Encounter with Race in the Twentieth-Century Urban North*. Chicago: University of Chicago Press, 1996.

———. "Thinking on One's Own: Catholicism in the American Intellectual Imagination, 1928–1960." *Journal of American History* 84, no. 1 (June 1997): 97–131.

McInerny, Ralph M. *The Defamation of Pius XII*. South Bend, IN: St. Augustine's Press, 2001.

McKeown, Elizabeth K. "Catholic Identity in America." In *America in Theological Perspective*, edited by Thomas M. McFadden. New York: Seabury Press, 1976.

———. "From *Pascendi* to *Primitive Man*: The Apologetics and Anthropology of John Montgomery Cooper." *US Catholic Historian* 13, no. 2 (Winter 1995): 1–22.

———. *War and Welfare: American Catholics and World War I*. New York: Garland Press, 1988.

McShane, Joseph M., S.J. *"Sufficiently Radical": Catholicism, Progressivism, and the Bishops' Program of 1919*. Washington, DC: Catholic University of America Press, 1986.

Mehler, Barry. "A History of the American Eugenics Society, 1921–1940." PhD diss., University of Illinois at Urbana-Champaign, 1988.

Neenan, Benedict. *Thomas Verner Moore: Psychiatrist, Educator, and Monk*. New York: Paulist Press, 2000.

Newman, Louise Michele. *White Women's Rights: The Racial Origins of Feminism in the United States*. New York: Oxford University Press, 1999.

Noble, David W. *Death of a Nation: American Culture and the End of Exceptionalism*. Minneapolis: University of Minnesota Press, 2002.

Noll, Steven. *Feeble-Minded in Our Midst: Institutions for the Mentally Retarded in the South, 1900–1940*. Chapel Hill: University of North Carolina Press, 1995.

Noonan, John T. *Contraception: A History of Its Treatment by the Catholic Theologians and Canonists*. 1965. Reprint, New York: Mentor-Omega, New American Library, 1967.

Oates, Mary J. *The Catholic Philanthropic Tradition in America*. Bloomington: Indiana University Press, 1995.

O'Brien, David J. *American Catholics and Social Reform: The New Deal Years*. New York: Oxford University Press, 1968.

O'Halloran, Ruth. "Organized Catholic Laywomen: The National Council of Catholic Women, 1920–1995." PhD diss., Catholic University of America, 1995.

O'Leary, Don. *Roman Catholicism and Modern Science: A History*. New York: Continuum, 2006.

Omi, Michael, and Howard Winant. *Racial Formation in the United States: From the 1960s to the 1990s*. 2nd ed. New York: Routledge, 1994.

Orsi, Robert A. *The Madonna of 115th Street: Faith and Community in Italian Harlem*. New Haven, CT: Yale University Press, 1985.

Pascoe, Peggy. "Miscegenation Law, Court Cases, and Ideologies of 'Race' in Twentieth-Century America." *Journal of American History* 83, no. 1 (June 1996): 44–69.

———. *What Comes Naturally: Miscegenation Law and the Making of Race in America*. New York: Oxford University Press, 2009.

Passelecq, George, and Bernard Suchecky. *The Hidden Encyclical of Pius XI*. Translated by Steven Rendell. New York: Harcourt Brace, 1997.

Paul, Diane B. *Controlling Human Heredity, 1865 to the Present*. Amherst, NY: Humanity Books, 1998.

———. "Eugenics and the Left." *Journal of the History of Ideas* 45, no. 4 (October–December 1984): 567–90.

Phayer, Michael. *The Catholic Church and the Holocaust, 1930–1965*. Bloomington: Indiana University Press, 2000.

Pickens, Donald K. *Eugenics and the Progressives*. Nashville, TN: Vanderbilt University Press, 1968.

Pivar, David. *Purity Crusade: Sexual Morality and Social Control, 1868–1900*. Westport, CT: Greenwood Press, 1973.

Poluse, Martin Frank. "Archbishop Joseph Schrembs and the Twentieth Century Catholic Church in Cleveland, 1921–1945." PhD diss., Kent State University, 1991.

"Popenoe, Paul (Bowman)." In *Contemporary Authors: A Bio-Bibliographical Guide to Current Writers in Fiction, General Nonfiction, Poetry, Journalism, Drama, Motion Pictures,*

Televisions, and Other Fields, 27:381. (New Revision Series, edited by Hal May and James G. Lesniak). Detroit: Gale Research, 1989.

Porter, Theodore M. "Fisher, Irving." In *American National Biography Online*. Oxford University Press, 2000–2010. http://www.anb.org/articles/14/14-00190.html.

Rafter, Nicole Hahn., ed. *White Trash: The Eugenic Family, 1877–1919*. Boston, MA: Northeastern University Press, 1986.

Rapp, Rayna. *Testing Women, Testing the Fetus: The Social Impact of Amniocentesis in America*. New York: Routledge, 1999.

Reed, James. *The Birth Control Movement and American Society: From Private Vice to Public Virtue*. Princeton, NJ: Princeton University Press, 1978.

Reilly, Philip R. "Laughlin, Harry Hamilton." In *American National Biography Online*. Oxford University Press, 2000–2010. http://www.anb.org/articles/13/13-00958.html.

———. *The Surgical Solution: A History of Involuntary Sterilization in the United States*. Baltimore: Johns Hopkins University Press, 1991.

Roberts, Dorothy. *Killing the Black Body: Race, Reproduction, and the Meaning of Liberty*. New York: Vintage Books, 1997.

Rosen, Christine. *Preaching Eugenics: Religious Leaders and the American Eugenics Movement*. Oxford: Oxford University Press, 2004.

Ross, Dorothy. *The Origins of American Social Science*. Cambridge: Cambridge University Press, 1992.

Rossiter, Margaret W. *Women Scientists in America: Struggles and Strategies to 1940*. Baltimore: Johns Hopkins University Press, 1982.

Russett, Cynthia Eagle. *Darwin in America: The Intellectual Response, 1865–1912*. San Francisco: W. H. Freeman, 1976.

Sánchez, José M. *Pius XII and the Holocaust: Understanding the Controversy*. Washington, DC: Catholic University of America Press, 2002.

Schoen, Johanna. *Choice and Coercion: Birth Control, Sterilization, and Abortion in Public Health and Welfare*. Chapel Hill: University of North Carolina, 2005.

"Scholasticism." In *New Catholic Encyclopedia*, 12:1153–70. Washington, DC: Catholic University Press, 1967.

Selden, Steven. *Inheriting Shame: The Story of Eugenics and Racism in America*. New York: Teachers College Press, 1999.

Shanabruch, Charles. *Chicago Catholics: The Evolution of an American Identity*. Notre Dame, IN: University of Notre Dame Press, 1981.

Shelly, Thomas J. "The Oregon School Case and the National Catholic Welfare Conference." *Catholic Historical Review* 75, no. 3 (1989): 439–57.

Sherman, Richard B. "'The Last Stand': The Fight for Racial Integrity in Virginia in the 1920s." *Journal of Southern History* 54, no. 1 (1988): 69–92.

Slawson, Douglas J. *The Department of Education Battle, 1918–1932: Public Schools, Catholic Schools, and the Social Order*. Notre Dame, IN: University of Notre Dame Press, 2005.

———. *The Foundation and First Decade of the National Catholic Welfare Council*. Washington, DC: Catholic University of America Press, 1992.

Smith, J. David, and K. Ray Nelson. *The Sterilization of Carrie Buck*. Far Hills, NJ: New Horizon, 1989.

Southern, David W. "But Think of the Kids: Catholic Interracialists and the Great American Taboo of Race Mixing." *U.S. Catholic Historian* 16, no. 3 (Summer 1998): 67–93.

———. *John LaFarge and the Limits of Catholic Interracialism, 1911–1963*. Baton Rouge: Louisiana State University Press, 1996.

Stepan, Nancy Leys. *"The Hour of Eugenics": Race, Gender, and Nation in Latin America.* Ithaca, NY: Cornell University Press, 1996.

Stepan, Nancy Leys, and Sander L. Gilman. "Appropriating the Idioms of Science: The Rejection of Scientific Racism." In *The Bounds of Race: Perspectives on Hegemony and Resistance,* edited by Dominick LaCapra, 73–103. Ithaca, NY: Cornell University Press, 1991.

Stern, Alexandra Minna. *Eugenic Nation: Faults and Frontiers of Better Breeding in Modern America.* Berkeley: University of California Press, 2005.

Stolba, Christine Alison. "'A Corrupt Tree Bringeth Forth Evil Fruit': Religion and the American Eugenics Movement." PhD diss., Emory University, 1999.

Tentler, Leslie Woodcock. "'The Abominable Crime of Onan': Catholic Pastoral Practice and Family Limitation, 1875–1919." *Church History* 71, no. 2 (June 2002): 307–40.

———. *Catholics and Contraception: An American History.* Ithaca, NY: Cornell University Press, 2004.

———. "On the Margins: The State of American Catholic History." *American Quarterly* 45, no. 1 (March 1993): 104–27.

Terry, Jennifer. "'Momism' and the Making of Treasonous Homosexuals." In *"Bad" Mothers: The Politics of Blame in Twentieth-Century America,* edited by Molly Ladd-Taylor and Lauri Umansky, 169–90. New York: New York University Press, 1998.

Tobin-Schlesinger, Kathleen. "Population and Power: The Religious Debate over Contraception, 1916–1936." 2 vols. PhD diss., University of Chicago, 1994.

Tone, Andrea. *Devices and Desires: A History of Contraceptives in America.* New York: Hill and Wang, 2001.

Tracy, David. *The Analogical Imagination: Christian Theology and the Culture of Pluralism.* New York: Crossroads, 1981.

Trent, James W. *Inventing the Feeble Mind: A History of Mental Retardation in the United States.* Berkeley: University of California Press, 1994.

VanEssendelft, William Ray. "A History of the Association for Voluntary Sterilization: 1935–1964." PhD diss., University of Minnesota, 1978.

White, Ronald C., Jr., and C. Howard Hopkins, eds. *The Social Gospel: Religion and Reform in Changing America.* Philadelphia: Temple University Press, 1976.

Wiebe, Robert H. *The Search for Order, 1877–1920.* New York: Hill and Wang, 1967.

Williamson, Joel. *New People: Miscegenation and Mulattos in the United States.* Baton Rouge: Louisiana State Press, 1995.

Wills, Garry. *Papal Sin: Structures of Deceit.* New York: Doubleday, 2000.

INDEX